Next Generation Graphene Nanomaterials for Cancer Theranostic Applications

Rohit Srivastava · Mukeshchand Thakur ·
Mukesh Kumar Kumawat ·
Rohan Bahadur

Next Generation Graphene Nanomaterials for Cancer Theranostic Applications

Springer

Rohit Srivastava
Department of Biosciences and Bioengineering
Indian Institute of Technology Bombay
Mumbai, Maharashtra, India

Mukeshchand Thakur
Department of Biosciences and Bioengineering
Indian Institute of Technology Bombay
Mumbai, Maharashtra, India

Mukesh Kumar Kumawat
Department of Biosciences and Bioengineering
Indian Institute of Technology Bombay
Mumbai, Maharashtra, India

Rohan Bahadur
Department of Biosciences and Bioengineering
Indian Institute of Technology Bombay
Mumbai, Maharashtra, India

ISBN 978-981-33-6302-1 ISBN 978-981-33-6303-8 (eBook)
https://doi.org/10.1007/978-981-33-6303-8

This Springer imprint is published by the registered company Springer Nature Singapore Pte Ltd.
The registered company address is: 152 Beach Road, #21-01/04 Gateway East, Singapore 189721, Singapore

Preface

Cancer has been a long-time player claiming lives around the globe as we write this book. Unsurprisingly, there are major research efforts in science and technology toward combating this condition in the diagnosis and therapeutics sector. We have come a long way understanding the complex mechanisms of cancer with some success on the way. However, we are still far behind and need new-generation rigorous and novel strategies to treat 'more effectively.' Many such strategies come from different fields of science including materials science and a most recent subset of it being nanotechnology.

Recently, there has been a boom of nanomaterials for cancer therapies right from metallic nanoparticles of gold, magnetic nanoparticles, and inorganic nanomaterials such as silica and soft materials such as hydrogels. These efforts have demonstrated unprecedented approaches and applications one could achieve in biology. One such field is the field of graphene-based nanomaterials ever since the discovery of graphene. Subsequently, numerous new families of carbon nanostructures were added such as small quantum dots resembling graphene nanomaterials with exciting optical properties which were not seen before. Therefore, in this book, we touch upon such 'new-generation' graphene-based nanomaterials—graphene oxide, graphene quantum dots, carbon dots, polymer dots, and so on. We discuss various aspects running from their discovery, properties, modifications, and finally, their role in diagnosis and therapy of cancer 'theranostics.'

Thus, this book has some material scientists, chemists, physicists, and bioengineers. We hope that this book will provide emerging material scientists hoping to focus their research in bioapplications for cancer therapy. The book will also help new researchers interested in the rapidly emerging field of carbon nanotechnology and biomaterials. We are thankful to the Indian Institute of Technology Bombay, India (IIT Bombay), for providing us with the scientific resources for compiling this book.

Mumbai, India

Rohit Srivastava
Rohan Bahadur
Mukesh Kumar Kumawat
Mukeshchand Thakur

Contents

About the Authors

Dr. Rohit Srivastava is currently working as a Professor and Head at the Department of Biosciences and Bioengineering, Indian Institute of Technology (IIT), Bombay. After completing his Bachelor's in Electronics Engineering from VNIT Nagpur in 1999, he joined Tata Consultancy Services, SEEPZ, Mumbai for a year and then went on to do a M.Sc. and Ph.D. in Biomedical Engineering from Louisiana Tech University, Ruston, USA. His specialization lies in PoC diagnostic devices, biomedical microsystems devices (MEMS), nanoengineered biosensors, photothermal therapy in cancers and nanoengineered orthopedic applications. His lab has funded projects across all domains, from point of care diagnostic devices to biosensors to cancer nanotechnology to MEMS drug delivery devices. He has co-authored over 120 research articles in international journals, 100 conference proceedings, and filed more than 100 U.S. and Indian patents, copyrights, and trademarks. He has been instrumental in technologies such as SYNC, ToucHB, Ucheck, and CareMother that have been successfully commercialized in the Indian market through his student entrepreneurs.

Mukeshchand Thakur worked at the Department of Biosciences and Bioengineering at Nanobios lab, Indian Institute of Technology (IIT) Bombay, India. While working at IIT Bombay, his research interests focused on the synthesis of multi-fluorescent quantum dots for biomedical applications. He has also participated in the Taiwan International Internship Program and is a recipient of the Royal Society of Chemistry Award (2014) at International Conference on Nano Science and Technology, India. He has authored multiple research publications in the field of biosensing and serves as a reviewer for over 10 international journals.

Dr. Mukesh Kumar Kumawat pursued his Ph.D. at the Department of Biosciences and Bioengineering at Indian Institute of Technology (IIT) Bombay, India, where his research focused on the synthesis, characterization, and applications of graphene nanomaterials in bio-imaging, sensing, drug delivery, and photothermal therapy of tumor cells. He has also worked on the exfoliation of MoS2 nanosheets,

MoS2 quantum dots synthesis, and their biomedical applications. He has co-authored research articles on the synthesis of gold nanomaterial, characterization, and their application for sensing applications. He has co-authored 15 research publications and 6 patents/applications.

Rohan Bahadur worked at the Department of Biosciences and Bioengineering in IIT Bombay, India on 2D materials such as graphene, MoS2, MXenes and their 0D analogues for biomedical and biosensing applications. As part of his Masters in VIT University, Vellore, India he did a research project at ICMCB, CNRS, University of Bordeaux, France on the production of patchy particles as building blocks of limited valence with Professor Etienne Duguet. He has co-authored 5 research publications and has 3 patents/applications.

Chapter 1
Introduction and Brief History

1.1 Introduction

Carbon forms the basis of life. It has its vast abundance in naturally occurring as well as synthetic substances. Carbon was recognized first in the eighteenth century. Carbon is a unique element of group 14 of the periodic table which has an atomic number 6 and denoted by the symbol 'C.' There are naturally available three isotopes of carbon-^{12}C, ^{13}C, and ^{14}C. First two isotopes are stable and the third one is a radionuclide. Carbon can not only bind itself but it can also bind exceptionally with numerous elements of the periodic table in the different ways and that is the key of its huge contribution to synthetic-chemistry. Carbon is found in various elemental forms that have different structures and physicochemical properties hence called 'allotropes' (Hirsch 2010). Graphite and diamond are very well-known allotropes since years. Over the years, the property of carbon to bind in different ways gave curiosity and scope to the scientists to discover the possibilities of finding the existence of other carbon allotropes. Eventually, the efforts of the scientists resulted in the discovery of other allotropes such as fullerenes, carbon nanotubes, graphene, graphyne, and cyclo [18] carbon.

1.2 Discovery of Carbon Allotropes

The history of the discovery of the carbon allotropes has really been interesting. Some research groups contributed in predicting the structures of the carbon allotropes by theoretical work and computational studies while other groups consistently worked on their preparation and realized their existence. In many cases, even the properties of these carbon structures were predicted through computational studies. Table 1.1 is describing all crucial stepping stones that led to the discoveries of carbon allotropes.

R. Srivastava et al., *Next Generation Graphene Nanomaterials for Cancer Theranostic Applications*,
https://doi.org/10.1007/978-981-33-6303-8_1

Table 1.1 A tabulated account of journeys of discoveries of the carbon allotropes and other significant contributions by researchers for various carbon forms

Years	Carbon structure and allotrope	Researchers	Contribution	References
1859	Graphite oxide	Benjamin Brodie	Oxidation of graphite	Brodie (1859)
1948	Graphite oxide	G. Ruess, F. Vogt	TEM observation of graphene oxide	Ruess and Vogt (1948)
1962	Reduced graphene oxide	Hofmann and Boehm	TEM observation of reduced graphene oxide and reported the presence of a few monolayers of graphene	Boehm et al. (1962)
1966	Fullerene	David Jones	Conceived the idea of having closed spheroidal cages made up of graphene sheets (with all hexagons)	Jones (1966)
1970	Fullerene	E. G. Osawa	Perceived the idea of having carbon atoms on the vertices of a truncated icosahedron and in that it would be stable aromatic structure	Osawa (1970)
1973	Fullerene	D. A. Bochvar, E. G. Gal'pern	Conceptualized that C_{60} as a closed shell molecule with a very large HOMO-LUMO gap	Bochvar and Galpern (1973)
1982	Fullerene	David Jones	Suggested that introducing pentagons in the graphene sheet could lead a closed cage molecule	Jones (1993)
1984	Carbon structures	I. V. Stankevich, M. V. Nikerov, D. A. Bochvar	Performed Huckel calculations	Stankevich et al. (1984)
1984	Carbon cluster	A. Kaldor, E. A. Rohlfing, and D. M. Cox	Studied carbon cluster formation on a reforming catalyst. A mass spectrum that first indicated the even numbered carbon clusters	Rohlfing et al. (1984)

(continued)

Table 1.1 (continued)

Years	Carbon structure and allotrope	Researchers	Contribution	References
1985	C_{60} fullerenes	H. W. Kroto, J. R. Heath, S. C. O'Brien, R. F. Curl and R. E. Smalley	Development of laser-vaporization supersonic cluster beam technique and microscope, nozzle design, and discovery of C_{60} fullerenes	Kroto et al. (1985)
1986	Nomenclature of graphite intercalation compounds	H. P. Boehm, R. Setton, and E. Stumpp	Coined the term 'Graphene'	Boehm et al. (1986)
1990	Graphene thin layers	K. Seibert, G. C. Cho, W. Kutt, and H. Kurz	Preparation of the graphene thin layers sample ($\sim$35 nm) by peeling-off from graphite using a Scotch tape. Reflectivity and transmission measurements	Seibert et al. (1990)
1991	Single-walled carbon nanotubes	S. Iijima	Synthesis of tubular carbon structures using arc-discharge method	Iijima (1991)
2001	Nanographene	A. M. Affoune, B. L. V. Prasad, Hirohiko Sato, Toshiaki Enoki, Yutaka Kaburagi, and Yoshihiro Hishiyama	Prepared a nanographene on a highly oriented pyrolytic graphite (HOPG) substrate	Affoune et al. (2001)
2004	Graphene	K. S. Novoselov, A. K. Geim, S. V. Morozov, D. Jiang, Y. Zhang, S. V. Dubonos, I. V. Grigorieva, A. A. Firsov	Separated graphene layers up to a few atomic thickness from graphite using a Scotch tape, fabricated a device, studied electronic properties in details	Novoselov et al. (2004)
2010	Polyynes (allotrope carbyne)	Wesley A. Chalifoux and Rik R. Tykwinski	Synthesized a series of conjugated polyynes as models for sp-hybridized carbon allotrope carbyne	Chalifoux and Tykwinski (2010)
2019	Cyclo [18] carbon	Katharina Kaiser1, Lorel M. Scriven, Fabian Schulz, Przemyslaw Gawel, Leo Gross, and Harry L. Anderson	Synthesis, isolation, and characterization of cyclo [18] carbon	Kaiser et al. (2019)

The discovery of fullerenes has been a wonderful adventure over the years (Smalley 1997) (see Table 1.1). Many thousands years ago, Archimedes proposed the idea of truncated icosahedral geometry. In 1970, E. G. Osawa conceived that carbon atoms at the vertices of an icosahedron could make a stable aromatic structure (Osawa 1970). After 1970 onward, up to more than a decade, the stability of the truncated icosahedron and its Huckel calculations were conceptualized. But, the process for bringing the structure into its existence was yet unknown.

Carbon, unlike other refractory elements, for example, such as platinum, tungsten, and tantalum, has an ability to cluster (C_n) when it is heated at high temperatures (3000–4000 K) in the gas phase. Contrarily, every element of the periodic table (except carbon) is found in monoatomic or diatomic state at above 1000 K. It is cohesive energy that is a responsible factor for catenation. Before 1984, the synthesis of the linear clusters was only reported and there was no speculation or whatsoever about the further growth of the clusters (Smalley 1997). It was unknown that what happens if the saturated vapors of carbon is allowed to condense to make a cluster with atomic range 40–100 and the technique that would enable this study was yet to be discovered. In 1980–1981, scientists at Rice University developed the laser-vaporization supersonic cluster beam technique that helped in studying the growth of the clusters from the high refractive metals and semiconductor elements from the periodic table (Michalopoulos et al. 1982; Hopkins et al. 1983). Professor Smalley's group at the Rice University constantly worked on developing the associated microscopic techniques too that could largely help them to view the nanoscopic aggregates of semiconductor and metal clusters with a range of 2–200 atoms. However, other scientists at Exxon, Andrew Kaldor's group were first who put carbon under such microscope for studying the carbon cluster formation on reforming catalyst (Rohlfing et al. 1984). While carrying out the study, they produced a mass spectrum that showed the presence of unprecedented carbon clusters with varying atomic strengths crossing over 100 atoms (Fig. 1.1). Interestingly, the clusters with atoms from 30 s till 150 atoms were found to be even numbered atomic strength which was never observed and reported before in the case of cluster formation by any other element from periodic table. The even numbered clusters indicated the presence of fullerenes and clusters before 30 s are linear and cyclic structures. The peak at C = 60 atoms was quite intense and distinguishable that indicated C_{60}. The prominence of the C_{60} peak could draw the attention of the Kaldor's group too but they didn't focus on the instrument components that their modification could play a pivotal role in separating the C_{60} from the mixture of other even numbered fullerenes. Professor Smalley's group at Rice University in collaboration with other researchers namely H. W. Croto, R. F. Curl concentrated on the nozzle conditions, cluster chemistry, and reactivity studies of the even numbered clusters. The supersonic nozzle design has two important components: (1) rotating disk and (2) integrating cup (see Fig. 1.2a). The integrating cup is an important component of the apparatus that is composed of nozzle and downstream flow restrictors. Cluster cooking reactions occur in the integrating cup and it facilitates the increasing of the C_{60} population by 50 times.

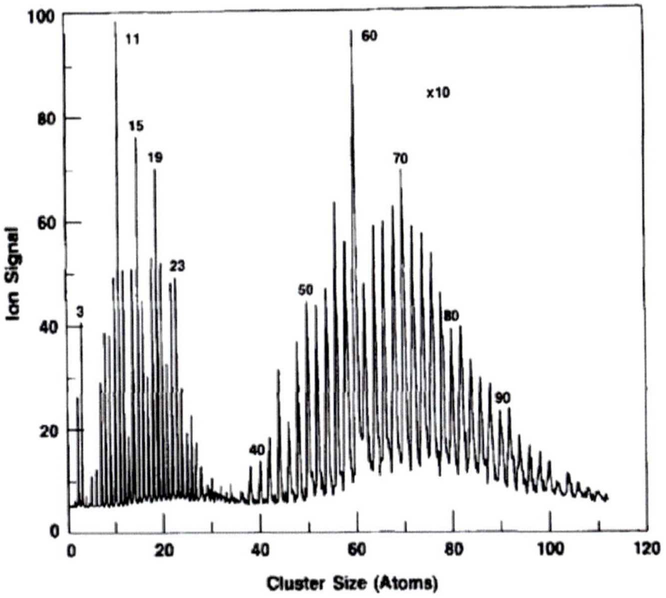

Fig. 1.1 A mass spectrum of the carbon clusters produced by Professor Kaldor's group. Adapted with permission from: Rohlfing et al. (1984)

The discovery of carbon nanotubes (CNTs) was done by Ijima in 1991 using arc-discharge method (Iijima 1991). However, meanwhile several researchers synthesized other carbon nanostructures (graphite whiskers, graphite filaments, carbon fibers, multi-walled carbon nanotubes, etc.) using arc-discharge and CVD methods (Bacon 1960; Bollmann and Spreadborough 1960; Lieberman et al. 1971; Oberlin et al. 1976). The CNTs can be called as rolled-up form of graphite sheets that can have length up to millimeters and diameter up to 2 nm. The discovery of next allotrope of carbon, i.e., graphene, was truly eureka moment. The term graphene was coined by Boehm and co-workers in 1986 (Boehm et al. 1986). The initial work on oxidation of graphite was done by Brodie in 1859 (Brodie 1859). He treated graphite with strong acids and obtained a yellow dispersion of graphite oxide which he believed to have discovered a new form of carbon 'graphon'. Graphon, discovered by Brodie, was actually graphene oxide, an oxidized form of graphite. Later, up to the next 100 years, no crucial developments were made in the discovery of graphene until the TEM studies of graphene oxide were carried out by Ruess and Vogt in 1948 (Ruess and Vogt 1948). Later, Hofmann and Boehm continued the TEM studies and claimed to have found a few monolayers of graphene while imaging the reduced graphene oxide sample (Boehm et al. 1962). However, the monolayers were not in the pure form and were present in the residue. That was the first observation of the monolayer graphene and it was based on the relative TEM contrast rather than focusing conditions. After 40 years, a few reports

Fig. 1.2 **a**. A diagram depicting the structure of the pulsed supersonic nozzle. Adapted with permission from: Kroto et al. (1985). **b**. Photograph of Professor R. E. Smalley holding the supersonic laser-vaporization cluster beam apparatus component that ultimately helped them in the discovery of C_{60} fullerenes. Adapted with permission from: Smalley (1997)

found the monolayers of graphene distinguishably by counting the folding lines under TEM (Horiuchi et al. 2004; Shioyama 2001; Viculis et al. 2003).

The graphite sheets could not be peeled-off up to single layer until 2004. In case of other metals, it was impossible to achieve a continuous monolayer of metals. However, ultrathin films of other metals were studied for many decades. When a few nanometers thick thin film is heated to achieve a monolayer, atoms coagulate and form tiny islands to yield a discontinuous film (Evans et al. 2006; Hopkins et al. 1983). In terms of carbon, the structures with three-dimensional configuration were found to be stable as compared to two-dimensional structures. Theoretically, it was stated that graphene was thermodynamically unstable, or least stable until the strength is about 6000 atoms (Shenderova et al. 2002). However, the efforts were made on growing graphene on a substrate (metal, insulating carbides, graphite, etc.) (Affoune et al. 2001; Blakely et al. 1970; Grant and Haas 1970; Nagashima et al. 1993; Van Bommel et al. 1975). In 2001, Hishiyama and co-workers prepared nanographene on a highly oriented pyrolytic graphite (HOPG) substrate using heat treatment at 1600 °C and electrophoretic deposition (Affoune et al. 2001). The research works reported about growing graphene films did not talk much about the film's quality and continuity. Another method for achieving graphene has been known as isolation of graphene sheets from graphite by peeling it off. In 1990, Kurz's group prepared a sample of graphene thin layers by peeling from pyrolytic graphite for reflectivity and transmission measurements using a Scotch tape (Seibert et al. 1990). In 2004, a team of researchers headed by A. K. Geim and K. S. Novoselov, successfully achieved high-quality graphene sheets having a few-layer to single-layer graphene sheets (Novoselov et al. 2004). They fabricated a device and studied electronic properties. They studied electric field effect and found that the resistivity changed by thousands times more than the previously reports about the metallic system. Apart from graphene, there are other synthetic allotropes that have been reported. These synthetic allotropes are a product of combination of differently hybridized carbon atoms (sp:carbyne, sp-sp^2: graphyne, sp-sp^3-yne-diamond, etc.,) graphyne is one well-known allotrope of them (Diederich and Kivala 2010; Hirsch 2010; Chalifoux and Tykwinski 2010).

1.3 Classification of Carbon Allotropes

The classification of carbon allotropes can be given on the following bases.

1. Classification based on availability;
2. Classification based on crystallinity;
3. Classification based on dimensionality;
4. Classification based on valence orbital hybridization.

1.3.1 Classification Based on Availability

There are several allotropes of carbon, some are naturally found and others have been predicted or/and synthesized in the laboratory (see Fig. 1.3).

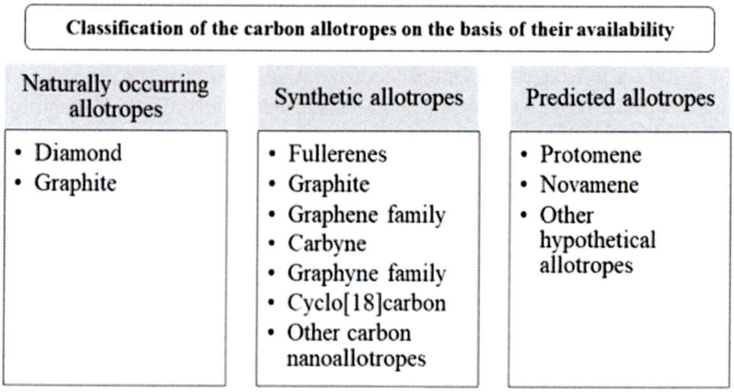

Fig. 1.3 Classification of carbon allotropes on the basis of their availability

1.3.2 Classification Based on Crystallinity

The classification of the carbon allotropes on the basis of crystallinity is given in Fig. 1.4. However, the crystallinity of carbon structures such as carbon nanofibers and graphene nanoribbons varies due to several parameters (Georgakilas et al. 2015).

1.3.3 Classification Based on Dimensionality

Carbon atoms in the different carbon allotropes are found in different hybridization states and eventually that form the structures with different morphologies. Figure 1.5 is depicting the carbon allotropes with different dimensionality

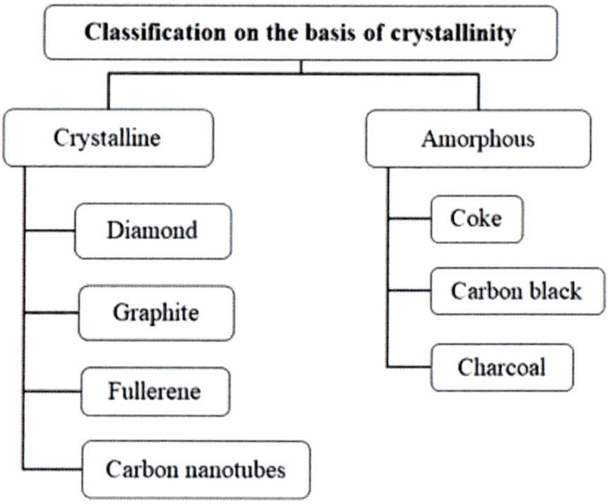

Fig. 1.4 Classification of carbon allotropes on the basis of crystallinity

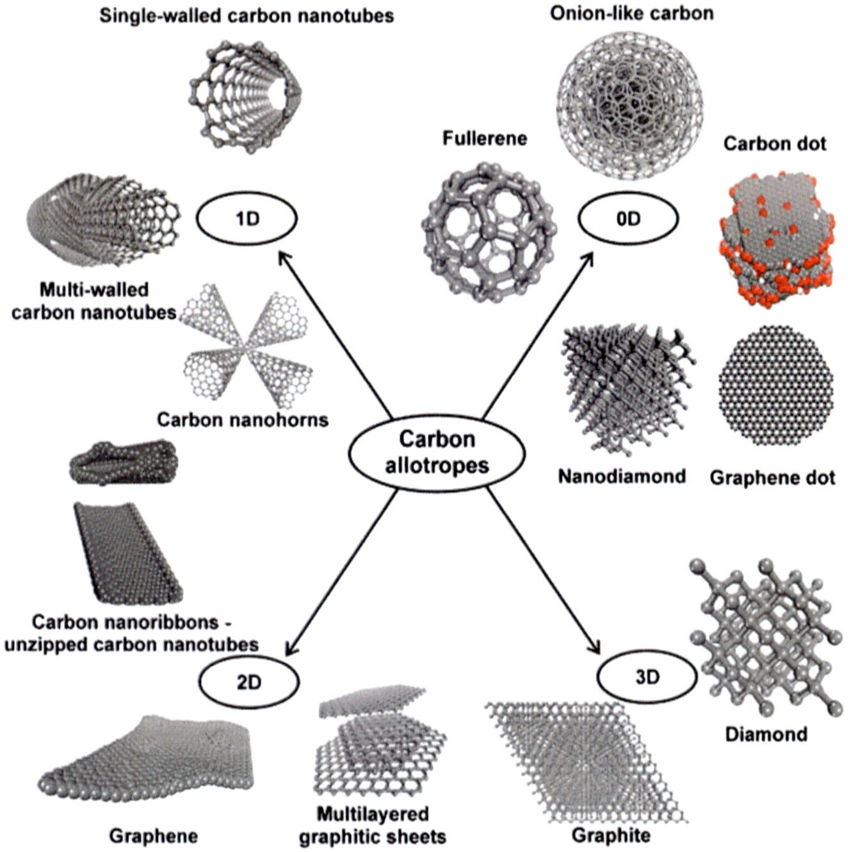

Fig. 1.5 Classification of carbon allotropes on the basis of dimensionality. Adapted with permission from: Georgakilas et al. (2015)

(Georgakilas et al. 2015). Fullerenes, carbon nano onions, carbon dots, nanodiamonds, and graphene dots are considered to be zero-dimensional (0D) allotropes. However, carbon nanotubes (single-walled and multi-walled) and carbon nanohorns are one-dimensional (1D) allotropes; single-atomic graphitic layer graphene, carbon nanoribbons, and unzipped carbon nanotubes are two-dimensional (2D) allotropes; and graphite and diamond are three-dimensional (3D) allotropes.

1.3.4 Classification Based on Valence Orbital Hybridization

The atoms in the carbon allotropes have different hybridization states. Some of them have purely single hybridization states while others have atoms with two different kinds of hybridization states. Figure 1.6 showing a ternary phase diagram

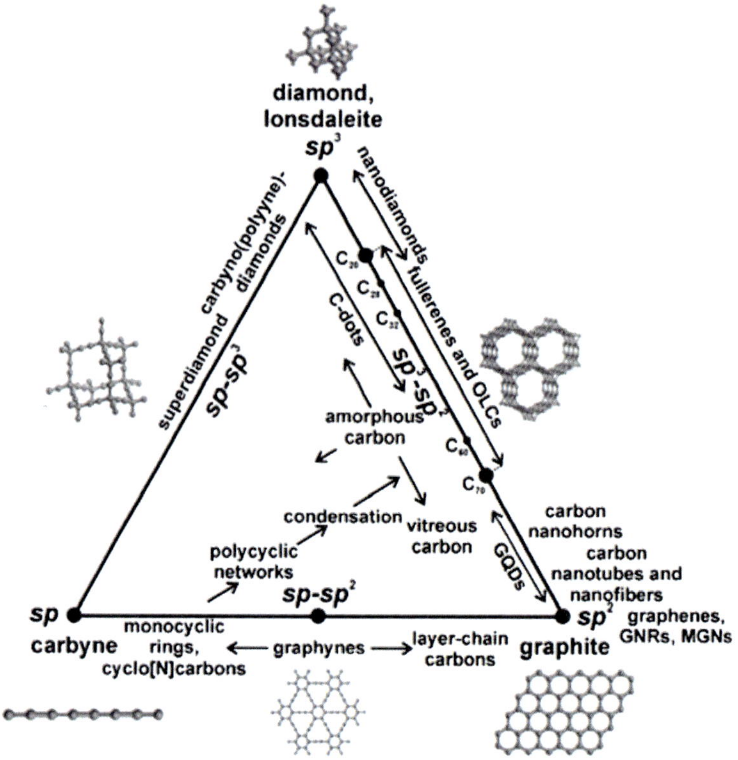

Fig. 1.6 Classification of carbon allotropes on the basis of valence orbital hybridization. Adapted with permission from: Heimann et al. (1997)

showing a distribution of carbon allotropes based on the hybridization states of the atoms in the allotropes (Heimann et al. 1997). The vertices of the diagram indicate the allotropes that have atoms with single kind of hybridization state. For example, graphite (sp^2), diamond (sp^3), and carbyne (sp) are made up of purely single kind of hybridized atoms. The allotropes which are displayed on the edge, have a mixture of atoms of two kinds of hybridization states of respective vertices that are connected with that edge. For example, the edge which connects sp and sp^2 vertices includes the allotropes such as cyclo [N] carbons and graphynes.

1.4 Cancer: Introduction

Cancer is one of the leading causes of the mortalities occurred worldwide. According to the data shown on the webpage of the World Health Organization (WHO), cancer has caused 9.6 million deaths worldwide in 2018 and it makes almost 16.6% of all mortalities (2018). Among all cancer types, lung and breast

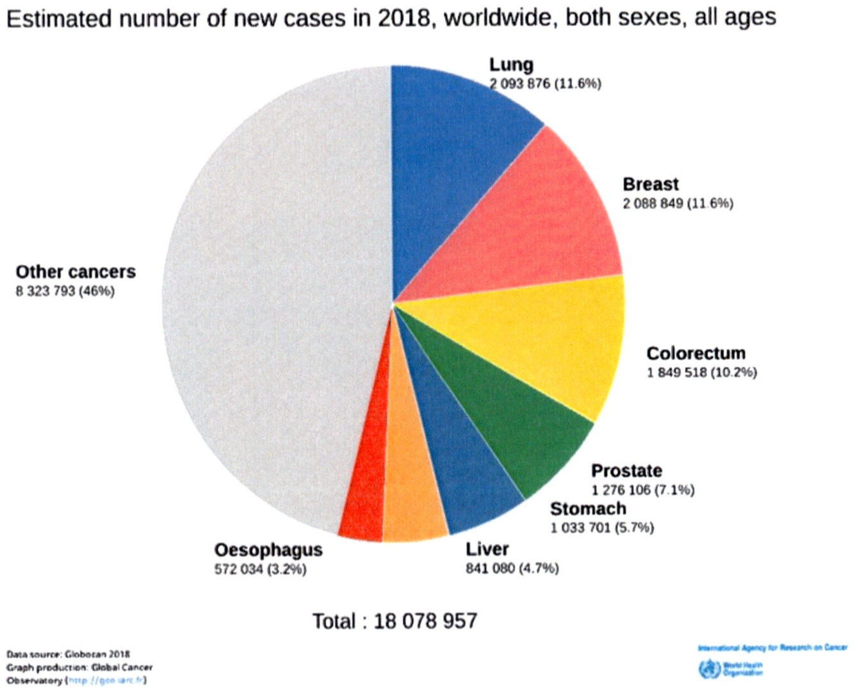

Fig. 1.7 New cases of cancer reported worldwide in 2018, among all age groups and both sexes. The figure has been accessed from webpage of World Health Organization; Link: https://bit.ly/3d51dRL (2018)

cancers cases are most abundant as per reports of 2018 (Fig. 1.7). Cancer is a group of diseases which begins with cells dividing uncontrollably and forming tumors. It can spread gradually to other organs of the body and this process is called metastasizing. The normal cells transform into tumor cells and division of the cells is genetically controlled. When the responsible genetic material interacts with the external agents that are called carcinogens, the genes are mutated and that affects the cell division process. These external agents include biological (obesity, aging, infections from certain bacteria, viruses, parasites, etc.), physical (UV radiation), and chemical carcinogens (tobacco, aflatoxin, asbestos, arsenic, etc.).

China and India are the top two most populated countries of the world. However, China is biggest victim of the cancer with 2.86 million deaths in 2018 alone (Fig. 1.8). The occurrence of the cases of lung and breast cancer did not have much difference in 2018 (Fig. 1.7); however, the lung cancer proved deadliest and caused 1.76 million deaths alone including both sexes and all age groups (Fig. 1.9) (2018). The lung cancer caused more deaths to males (highest among other cancer types) than the females with 1,184,947 and 576,060 deaths, respectively. However, breast cancer was the main cause of deaths in the females as compared to other cancer types causing 626,679 deaths alone (2018).

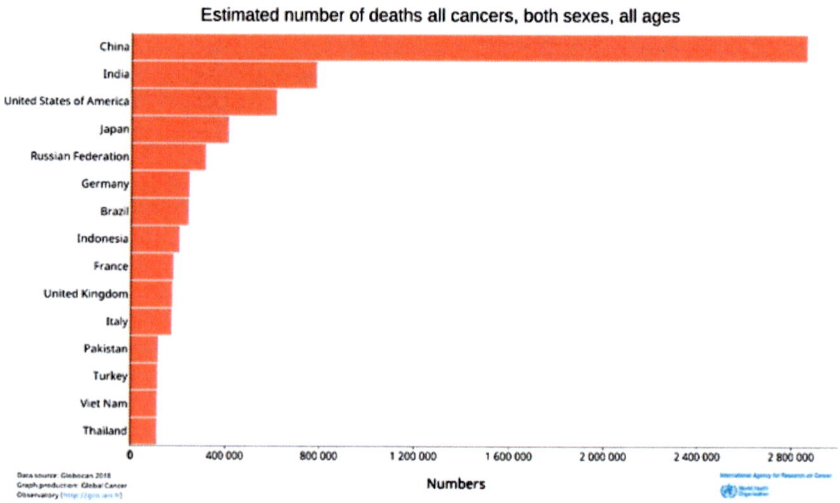

Fig. 1.8 Figure showing 15 leading counties in mortalities caused by cancer. The figure has been accessed from webpage of World Health Organization; Link: https://bit.ly/2XIrxKG

1.4.1 *Cancer: Worldwide Mortality*

See Figs. 1.8 and 1.9.

1.4.2 *Cancer Therapies: Current Scenario and Emergence of Cancer Nanomedicine*

There are three approaches in practice to treat the cancer: (1) chemotherapy, (2) radiotherapy, and (3) surgical excision. The preferences of these approaches depend on the cancer type and its stage. These therapies are either used separately or in combination. The chemotherapy is the major therapeutic approach among all. However, it has got several blockades that compromise its efficiency: (a) drug solubility, (b) multi-drug resistance, and (c) selectivity (Wicki et al. 2015).

(a) *Drug solubility*: Most of the chemotherapeutic drugs are hydrophobic in nature; hence, they require organic solvents for the dosage formulation. Using organic solvent and their subsequent traces might pose severe toxicity issue in the physiological medium (Kwon 2003).
(b) *Multi-drug resistance (MDR)*: It is one of the major reasons for the failure of chemotherapy. MDR is the resistance against the different kinds of anticancer drugs (Stavrovskaya 2000).
(c) *Selectivity*: Lack of selectivity in the anticancer drugs causes damage to neighboring healthy tissue (Wicki et al. 2015).

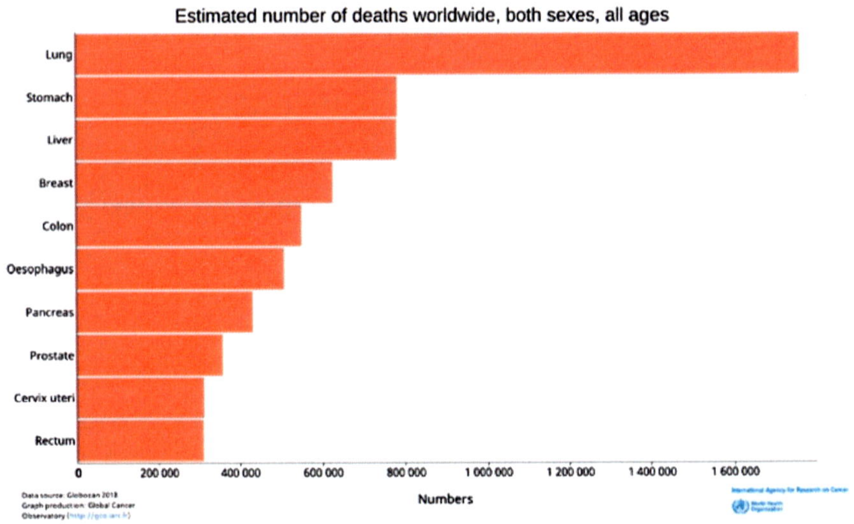

Fig. 1.9 Numbers showing deaths caused by different types of cancer worldwide in 2018. The figure has been accessed from webpage of World Health Organization; Link: https://bit.ly/39xTWbT

To overcome the drawbacks of the conventional therapeutic approaches, nanotechnology has brought a great revolution in the field of medicine. Nanotechnology for the medicine, i.e., nanomedicine, has brought several therapeutic modalities for cancer diagnosis, treatment, and management. Nanomedicine has excellent promising features that help in improving the performance of existing anticancer drugs (Fig. 1.10).

(a) *Bioavailability*: Nanotechnology-enabled nanoparticles (NPs) help in increasing the bioavailability of the hydrophobic drugs that have poor solubility in the aqueous medium. The drugs are encapsulated in the nanoparticles that provide additional protection to the drug and enhance its stability. For example, Wortmannin, Albumin-bound paclitaxel (Abraxane®), etc. Wortmannin, a radiosensitizer and PI3K inhibitor drug, was once abandoned due to its poor solubility and chemical instability, was coated with lipid-based nanocarriers, and had better performance (Karve et al. 2012; Reynolds et al. 2009).
(b) *Pharmacokinetic profile*: NPs help in improving the pharmacokinetic performance of the drug since NPs protect drug molecules from exposing toward enzymatic degradation and excretion.
(c) *Stimuli-sensitive release*: Some nanotherapeutics are designed in such a way that they release their payload or function upon some trigger such as pH and temperature.
(d) *Targeting*: Targeted nanomedicine facilitates a selective and more localized delivery of the drug molecules. It might decrease the drug resistance of the tumor cells. Nanocarriers can be functionalized with targeting ligands.

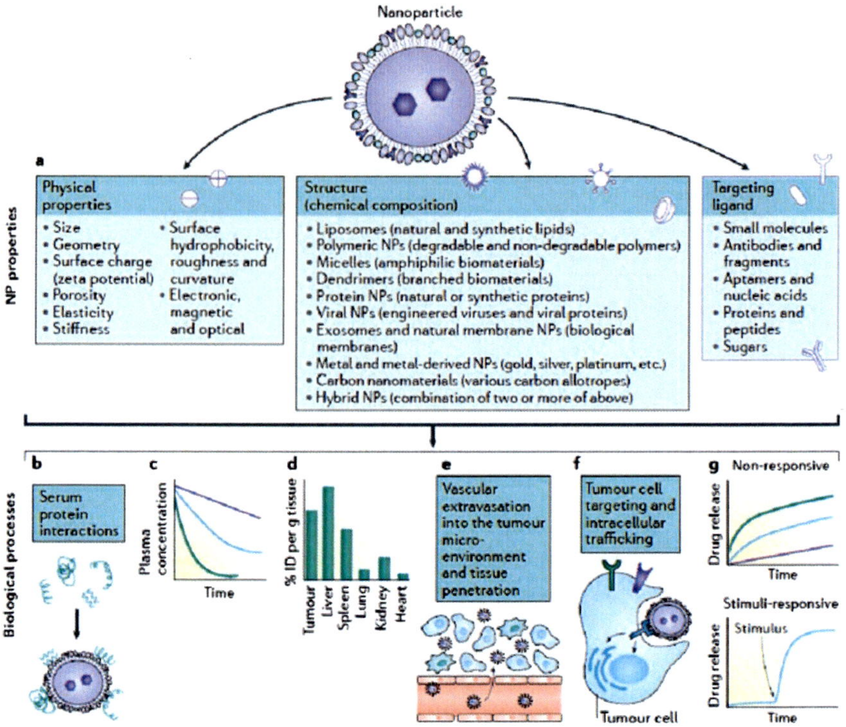

Fig. 1.10 Nanoparticle: structure, physical properties, and functionalization with targeting ligands. Effect of nanoparticle properties on its delivery to tumors

(e) *Improved circulation time*: Since the nanocarriers protect the drug molecule from enzymatic degradation and excretion, hence they increase the drug circulation (Wicki et al. 2015).

1.4.3 Cancer Nanomedicine: Recent Progress and Challenges

Cancer nanomedicine has achieved remarkable milestones in last few decades (Fig. 1.11) (Shi et al. 2017). Some notable achievements are mentioned in this section. A. D. Bangham and R. W. Horne studied liposome structure by negative staining of phospholipids and further observation under an electron microscope (Bangham and Horne 1964). The structure of the liposomes helped in predicting their possible applications. The early reports on liposomes being used as carrier for targeted delivery came in 1980 (Heath et al. 1980; Leserman et al. 1980). In 1964,

David M. Long and Judah Folkman reported that using a silicone rubber 'Silastic' as a carrier can help in prolonged delivery of the low molecular weight drug molecule in animal tissues (Folkman and Long 1964). Later, there have been a few reports that found limited success in sustain release of large molecular weight compounds by using polymers such as polyvinylpyrrolidone and polyacrylamide. But, the polymers mentioned in the report were causing inflammation in the tissues and could maintain the sustained release of the drug for a limited period (Davis 1972; Gimbrone et al. 1972). In 1976, Robert Langer and Judah Folkman prepared pellets using polymers such as Hydron, ethylene-vinyl acetate copolymer, and polyvinyl alcohol and demonstrated the release of large molecular weight compounds as long as 100 days (Langer and Folkman 1976). In 1986, Hiroshi Maeda and Yasuhiro Matsumura discovered enhanced permeability and retention (EPR) effect (Matsumura and Maeda 1986). They studied the distribution of 'Smancs,' a conjugate of copolymer styrene and maleic acid with an antitumor protein neocarzinostatin (NCS), in tumors and found that the smancs accumulated more effectively in the tumors than bare NCS. They further investigated the mechanism behind this phenomenon and speculated that this tumoritropic accumulation happened because of the leaky vasculature of the tumor tissue and poor drainage of the lymphatic system. This was a stepping stone toward macromolecular tumor treatment. A successful carrier needs to circulate in the blood for longer duration to deliver the drug at specific site. It is only possible when the carrier is not cleared out from the reticulo-endothelial system (RES). In 1994, Robert Langer and co-workers developed biodegradable nanospheres functionalized with polyethylene glycol (PEG). The nanospheres were composed of biodegradable polymers such as polycaprolactone (PCL), poly (lactic-co-glycolic acid) (PLGA), and their co-polymers. The covalent functionalization of the nanospheres with PEG helps in preventing the opsonization and subsequent recognition by the macrophages. This would ensure the longer circulation of the nanospheres in the blood stream and reduced uptake by the lever (Gref et al. 1994). The molecular weight (MW) of the PEG plays a crucial role, higher the MW make thicker the coating of PEG over nanosphere and increase the blood circulation time of the nanospheres. Later, in next two decades, significant research works were carried out in the field of nanomedicine. Various types of nanomaterials with different nature and composition such as organic, inorganic, organo-inorganic, protein, lipid, and synthetic polymers have been used for cancer therapeutic application (Fig. 1.12).

The nanotechnology enabled chemotherapy modalities that have marked immense achievements and have got approved by US Food and Drug Administration (FDA) or are under clinical trials include non-targeted delivery, targeted delivery, stimuli-responsive delivery, and combinatorial delivery. Other modalities include hyperthermia, radiotherapy, gene or RNAi therapy, and immunotherapy (Shi et al. 2017) (Table 1.2).

Nanomedicine is a promising approach though there are numerous challenges that nanomedicine has to face when it transits from laboratory to clinical stage and then clinical stage to market. Some of the challenges are as follows:

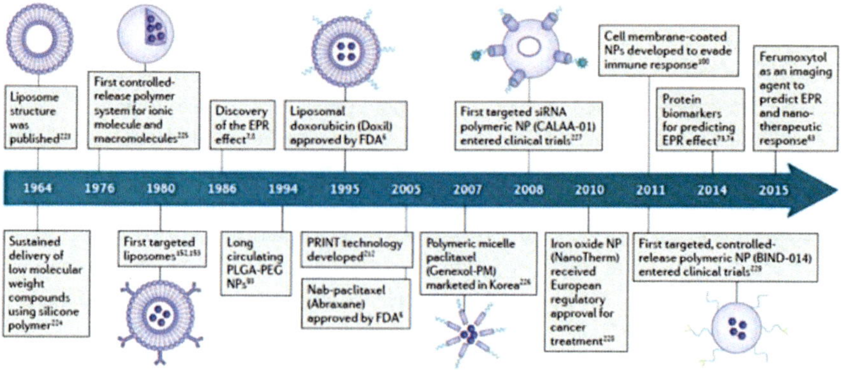

Fig. 1.11 A timeline showing major milestones achieved in the field of nanomedicine. Adapted with permission from: Shi et al. (2017)

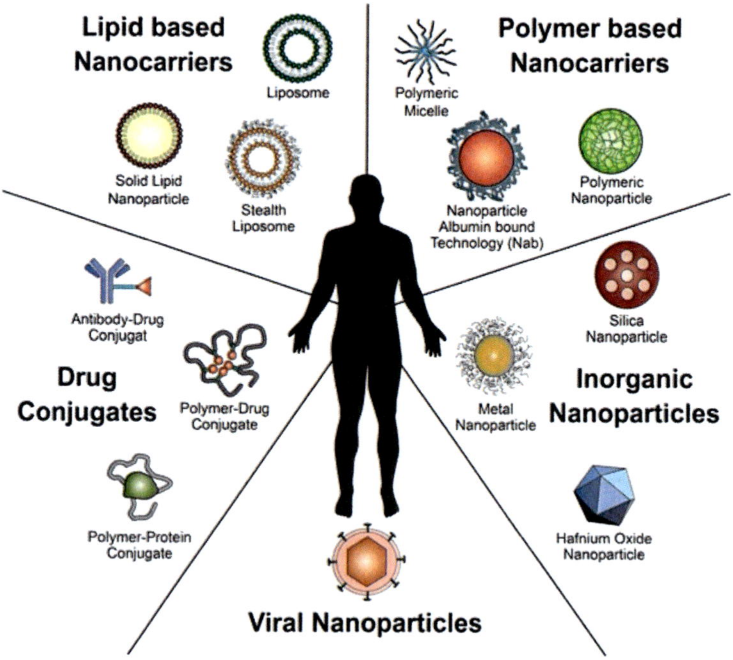

Fig. 1.12 An illustration of various nanotherapeutic platforms for clinical cancer care. Adapted with permission from: Wicki et al. (2015)

Table 1.2 Some examples of nanomedicines that are currently in market or under clinical trials. Contents of the table adapted with permission from: Shi et al. (2017), Wicki et al. (2015)

Product name	Material/conjugate	Active pharmaceutical ingredient	Cancer type	Status	References
Doxil	Pegylated liposome	Doxorubicin	Ovarian cancer, HIV-related Kaposi's sarcoma, and multiple myeloma	FDA-approved	Smith (2013)
Abraxane	Albumin-bound nanoparticle	Paclitaxel	Breast cancer, pancreatic cancer	FDA-approved	Desai et al. (2006)
DepoCyt	Liposome	Cytosine Arabinoside (Cytarabine)	Neoplastic meningitis	FDA-approved	Glantz et al. (1999)
DaunoXome	Liposome	Daunorubicin	Kaposi's sarcoma	FDA-approved	Rivera (2003)
Marqibo	Liposome	Voncristine	Acute lymphoid leukemia	FDA-approved	Rodriguez et al. (2009)
Onivyde or MM-398	Pegylated liposome	Irinotecan	Post-gemcitabine metastatic pancreatic cancer	FDA-approved	Inman (2015)
Lipo-Dox	Liposome	Doxorubicin	Breast and ovarian cancer, Kaposi's sarcoma	Approved in Taiwan	Khemapech et al. (2013)
Mepact	Liposome	Mifamurtide MTP-PE	Osteosarcoma	Approved in Europe	Venkatakrishnan et al. (2014)
Myocet	Liposome	Doxorubicin	Breast cancer	Approved in Europe	Batist et al. (2001)
Oncaspar	PEG-protein conjugate	L-asparaginase	Leukemia	FDA-approved	Dinndorf et al. (2007)
Genexol-PM	PEG-PLA polymeric micelle	Paclitaxel	Breast cancer, lung cancer, ovarian cancer	Approved in South Korea	Lee et al. (2008)

(continued)

Table 1.2 (continued)

Product name	Material/conjugate	Active pharmaceutical ingredient	Cancer type	Status	References
SMANCS	Polymer-protein conjugate	Neocarzinostatin	Liver cancer, renal cancer	Approved in Japan	Maeda (2001)
Lipoplatin	Pegylated liposome	Cisplatin	Non-small-cell lung carcinoma (NSCLC)	Phase-III	Stathopoulos et al. (2010)
NK-105	Polymeric micelle	Paclitaxel	Metastatic breast cancer	Phase-III	ClinicalTrials.gov (2016c)
ThermoDox	Liposome	Doxorubicin	Hepatocellular carcinoma	Phase-III	ClinicalTrials.gov (2016b)
AuroLase	Gold-silica nanoshell (Hyperthermia)	NA	Head and neck cancer, primary and metastatic lung tumors	Pilot study	ClinicalTrials.gov (2016a, 2015b)
NanoTherm	Iron oxide NPs (Hyperthermia)	NA	Glioblastoma	Approved in Europe	Smith (2013)
NBTXR3	Hafnium oxide NPs (Radiotherapy)	NA	Adult soft tissue sarcoma	Phase-II/III	ClinicalTrials.gov (2016d)
MRX34	Liposome	miR-34 mimic	Solid tumors, primary liver cancer, and haematological malignancies	Phase-I	ClinicalTrials.gov (2016e))
Tacemotide	Liposome	MUC1 antigen	NSCLC	Phase-III	ClinicalTrials.gov (2015c)
DPX-0907	Liposome	Multi-tumor associated antigens	Ovarian, breast, and prostate cancer	Phase-I	ClinicalTrials.gov (2015a)
Lipovaxin-MM	Liposome	Melanoma antigens	Malignant Melanoma	Phase-I	ClinicalTrials.gov (2012)

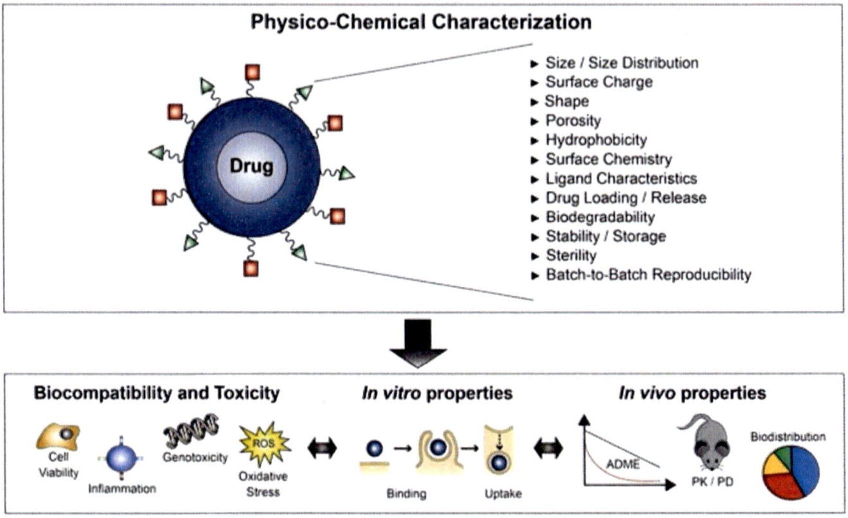

Fig. 1.13 Physicochemical parameters that are important in determining the in vitro and in vivo interactions of the nanomaterials. Adapted with permission from: Wicki et al. (2015)

- Physicochemical characterization and optimization: Physicochemical parameters are crucial for determination of success of the nanomedicine (Fig. 1.13). These parameters have much to control various important events such as cell targeting and internalization, toxicity, immune evasion, controlled drug release, tumor extravasation and diffusion (Wicki et al. 2015).
- Controlled synthesis, reproducibility, and scalable production.
- Conventional in vitro assay does not essentially emulate the complexity and physiological environment of biological tissues.
- Discrepancies between the results obtained from preclinical studies and clinical trials.
- Requirement of the developing the animal tumor models that can exactly match with the human tumor models in terms of heterogeneity and anatomical histology.
- Requirement of good manufacturing practice (GMP) labs.
- Regulatory issues.

References

2012. ClinicalTrials.gov. https://clinicaltrials.gov/ct2/show/NCT01052142?term.
2015a. ClinicalTrials.gov. https://clinicaltrials.gov/ct2/show/NCT01095848?term.
2015b. ClinicalTrials.gov. https://clinicaltrials.gov/ct2/show/NCT01679470?term.
2015c. ClinicalTrials.gov. https://clinicaltrials.gov/ct2/show/NCT00409188?term.

2016a. ClinicalTrials.gov. https://clinicaltrials.gov/ct2/show/NCT00848042?term.
2016b. ClinicalTrials.gov. https://clinicaltrials.gov/ct2/show/NCT02112656?term.
2016c. ClinicalTrials.gov. https://clinicaltrials.gov/ct2/show/NCT01644890?term.
2016d. ClinicalTrials.gov. https://clinicaltrials.gov/ct2/show/NCT02379845?term.
2016e. ClinicalTrials.gov. https://clinicaltrials.gov/ct2/show/NCT01829971?term.
2018. World Health Organization. Available https://gco.iarc.fr/. (Accessed).

Affoune, A., Prasad, B., Sato, H., Enoki, T., Kaburagi, Y., & Hishiyama, Y. (2001). Experimental evidence of a single nano-graphene. *Chemical Physics Letters, 348,* 17–20.

Bacon, R. (1960). Growth, structure, and properties of graphite whiskers. *Journal of Applied Physics, 31,* 283–290.

Bangham, A. D., & Horne, R. (1964). Negative staining of phospholipids and their structural modification by surface-active agents as observed in the electron microscope. *Journal of Molecular Biology, 8,* 660–IN10.

Batist, G., Ramakrishnan, G., Rao, C. S., Chandrasekharan, A., Gutheil, J., Guthrie, T., et al. (2001). Reduced cardiotoxicity and preserved antitumor efficacy of liposome-encapsulated doxorubicin and cyclophosphamide compared with conventional doxorubicin and cyclophosphamide in a randomized, multicenter trial of metastatic breast cancer. *Journal of Clinical Oncology, 19,* 1444–1454.

Blakely, J., Kim, J., & Potter, H. (1970). Segregation of carbon to the (100) surface of nickel. *Journal of Applied Physics, 41,* 2693–2697.

Bochvar, D., & Galpern, E. (1973). Hypothetical systems-carbododecahedron, s-icosahedrone and carbo-s-icosahedron. *Doklady Akademii Nauk Sssr, 209,* 610–612.

Boehm, H.-P., Clauss, A., Fischer, G., & Hofmann, U. (1962). Das adsorptionsverhalten sehr dünner kohlenstoff-folien. *Zeitschrift für anorganische und allgemeine Chemie, 316,* 119–127.

Boehm, H., Setton, R., & Stumpp, E. (1986). Nomenclature and terminology of graphite intercalation compounds. Pergamon.

Bollmann, W., & Spreadborough, J. (1960). Action of graphite as a lubricant. *Nature, 186,* 29–30.

Brodie, B. C. (1859). XIII. On the atomic weight of graphite. *Philosophical Transactions of the Royal Society of London*, 249–259.

Chalifoux, W. A., & Tykwinski, R. R. (2010). Synthesis of polyynes to model the sp-carbon allotrope carbyne. *Nature Chemistry, 2,* 967–971.

Davis, B. K. (1972). Control of diabetes with polyacrylamide implants containing insulin. *Experientia, 28,* 348.

Desai, N., Trieu, V., Yao, Z., Louie, L., Ci, S., Yang, A., et al. (2006). Increased antitumor activity, intratumor paclitaxel concentrations, and endothelial cell transport of cremophor-free, albumin-bound paclitaxel, ABI-007, compared with cremophor-based paclitaxel. *Clinical Cancer Research, 12,* 1317–1324.

Diederich, F., & Kivala, M. (2010). All-carbon scaffolds by rational design. *Advanced Materials, 22,* 803–812.

Dinndorf, P. A., Gootenberg, J., Cohen, M. H., Keegan, P., & Pazdur, R. (2007). FDA drug approval summary: Pegaspargase (oncaspar) for the first-line treatment of children with acute lymphoblastic leukemia (ALL). *Oncologist, 12,* 991–998.

Evans, J., Thiel, P., & Bartelt, M. C. (2006). Morphological evolution during epitaxial thin film growth: Formation of 2D islands and 3D mounds. *Surface Science Reports, 61,* 1–128.

Folkman, J., & Long, D. M. (1964). The use of silicone rubber as a carrier for prolonged drug therapy. *Journal of Surgical Research, 4,* 139–142.

Georgakilas, V., Perman, J. A., Tucek, J., & Zboril, R. (2015). Broad family of carbon nanoallotropes: Classification, chemistry, and applications of fullerenes, carbon dots, nanotubes, graphene, nanodiamonds, and combined superstructures. *Chemical Reviews, 115,* 4744–4822.

Gimbrone, M. A., Jr., Leapman, S. B., Cotran, R. S., & Folkman, J. (1972). Tumor dormancy in vivo by prevention of neovascularization. *Journal of Experimental Medicine, 136,* 261–276.

Glantz, M. J., Jaeckle, K. A., Chamberlain, M. C., Phuphanich, S., Recht, L., Swinnen, L. J., et al. (1999). A randomized controlled trial comparing intrathecal sustained-release cytarabine

(DepoCyt) to intrathecal methotrexate in patients with neoplastic meningitis from solid tumors. *Clinical Cancer Research, 5,* 3394–3402.
Grant, J., & Haas, T. (1970). A study of Ru (0001) and Rh (111) surfaces using LEED and Auger electron spectroscopy. *Surface Science, 21,* 76–85.
Gref, R., Minamitake, Y., Peracchia, M. T., Trubetskoy, V., Torchilin, V., & Langer, R. (1994). Biodegradable long-circulating polymeric nanospheres. *Science, 263,* 1600–1603.
Heath, T. D., Fraley, R. T., & Papahdjopoulos, D. (1980). Antibody targeting of liposomes: cell specificity obtained by conjugation of F (ab') 2 to vesicle surface. *Science, 210,* 539–541.
Heimann, R., Evsyukov, S., & Koga, Y. (1997). Carbon allotropes: A suggested classification scheme based on valence orbital hybridization. *Carbon (New York, NY), 35,* 1654–1658.
Hirsch, A. (2010). The era of carbon allotropes. *Nature Materials, 9,* 868–871.
Hopkins, J., Langridge-Smith, P., Morse, M., & Smalley, R. (1983). Supersonic metal cluster beams of refractory metals: Spectral investigations of ultracold Mo2. *The Journal of Chemical Physics, 78,* 1627–1637.
Horiuchi, S., Gotou, T., Fujiwara, M., Asaka, T., Yokosawa, T., & Matsui, Y. (2004). Single graphene sheet detected in a carbon nanofilm. *Applied Physics Letters, 84,* 2403–2405.
Iijima, S. (1991). Helical microtubules of graphitic carbon. *Nature, 354,* 56–58.
Inman, S. (2015). FDA approves second-line MM-398 regimen for metastatic pancreatic cancer. *OncLive.*
Jones, D. E. (1966). Hollow molecules. *New Sci, 32,* 245.
Jones, D. E. (1993). Dreams in a charcoal fire: Predictions about giant fullerenes and graphite nanotubes. *Philosophical Transactions of the Royal Society of London Series A: Physical and Engineering Sciences, 343,* 9–18.
Kaiser, K., Scriven, L. M., Schulz, F., Gawel, P., Gross, L., & Anderson, H. L. (2019). An sp-hybridized molecular carbon allotrope, cyclo [18] carbon. *Science, 365,* 1299–1301.
Karve, S., Werner, M. E., Sukumar, R., Cummings, N. D., Copp, J. A., Wang, E. C., et al. (2012). Revival of the abandoned therapeutic wortmannin by nanoparticle drug delivery. *Proceedings of the National Academy of Sciences U S A, 109,* 8230–8235.
Khemapech, N., Oranratanaphan, S., Termrungruanglert, W., Lertkhachonsuk, R., & Vasurattana, A. (2013). Salvage chemotherapy in recurrent platinum-resistant or refractory epithelial ovarian cancer with carboplatin and distearoylphosphatidylcholine pegylated liposomal doxorubicin (lipo-dox(R)). *Asian Pacific Journal of Cancer Prevention, 14,* 2131–2135.
Kroto, H. W., Heath, J. R., O'brien, S. C., Curl, R. F., & Smalley, R. E. (1985). C60: Buckminsterfullerene. *Nature, 318,* 162–163.
Kwon, G. S. (2003). Polymeric micelles for delivery of poorly water-soluble compounds. *Critical Reviews in Therapeutic Drug Carrier Systems, 20,* 357–403.
Langer, R., & Folkman, J. (1976). Polymers for the sustained release of proteins and other macromolecules. *Nature, 263,* 797–800.
Lee, K. S., Chung, H. C., Im, S. A., Park, Y. H., Kim, C. S., Kim, S. B., et al. (2008). Multicenter phase II trial of genexol-PM, a cremophor-free, polymeric micelle formulation of paclitaxel, in patients with metastatic breast cancer. *Breast Cancer Research and Treatment, 108,* 241–250.
Leserman, L. D., Barbet, J., Kourilsky, F., & Weinstein, J. N. (1980). Targeting to cells of fluorescent liposomes covalently coupled with monoclonal antibody or protein A. *Nature, 288,* 602–604.
Lieberman, M. L., Hills, C. R., & Miglionico, C. J. (1971). Growth of graphite filaments. *Carbon, 9,* 633–635.
Maeda, H. (2001). SMANCS and polymer-conjugated macromolecular drugs: Advantages in cancer chemotherapy. *Advanced Drug Delivery Reviews, 46,* 169–185.
Matsumura, Y., & Maeda, H. (1986). A new concept for macromolecular therapeutics in cancer chemotherapy: Mechanism of tumoritropic accumulation of proteins and the antitumor agent smancs. *Cancer Research, 46,* 6387–6392.
Michalopoulos, D., Geusic, M., Hansen, S., Powers, D., & Smalley, R. (1982). The bond length of chromium dimer. *The Journal of Physical Chemistry, 86,* 3914–3916.

Nagashima, A., Nuka, K., Satoh, K., Itoh, H., Ichinokawa, T., Oshima, C., et al. (1993). Electronic structure of monolayer graphite on some transition metal carbide surfaces. *Surface Science, 287,* 609–613.

Novoselov, K. S., Geim, A. K., Morozov, S. V., Jiang, D., Zhang, Y., Dubonos, S. V., et al. (2004). Electric field effect in atomically thin carbon films. *Science, 306,* 666–669.

Oberlin, A., Endo, M., & Koyama, T. (1976). Filamentous growth of carbon through benzene decomposition. *Journal of Crystal Growth, 32,* 335–349.

Osawa, E. (1970). Superaromaticity. *Kagaku, 25,* 854–863.

Reynolds, C., Barrera, D., Jotte, R., Spira, A. I., Weissman, C., Boehm, K. A., et al. (2009). Phase II trial of nanoparticle albumin-bound paclitaxel, carboplatin, and bevacizumab in first-line patients with advanced nonsquamous non-small cell lung cancer. *Journal of Thoracic Oncology, 4,* 1537–1543.

Rivera, E. (2003). Liposomal anthracyclines in metastatic breast cancer: Clinical update. *Oncologist, 8*(Suppl 2), 3–9.

Rodriguez, M. A., Pytlik, R., Kozak, T., Chhanabhai, M., Gascoyne, R., Lu, B., et al. (2009). Vincristine sulfate liposomes injection (Marqibo) in heavily pretreated patients with refractory aggressive non-Hodgkin lymphoma: Report of the pivotal phase 2 study. *Cancer, 115,* 3475–3482.

Rohlfing, E. A., Cox, D. M., & Kaldor, A. (1984). Production and characterization of supersonic carbon cluster beams. *The Journal of Chemical Physics, 81,* 3322–3330.

Ruess, G., & Vogt, F. (1948). Höchstlamellarer Kohlenstoff aus Graphitoxyhydroxyd. *Monatshefte für Chemie und verwandte Teile anderer Wissenschaften, 78,* 222–242.

Seibert, K., Cho, G. C., Kutt, W., Kurz, H., Reitze, D. H., Dadap, J. I., et al. (1990). Femtosecond carrier dynamics in graphite. *Physical Review B: Condensed Matter, 42,* 2842–2851.

Shenderova, O., Zhirnov, V., & Brenner, D. (2002). Solid State Materials. *Science, 27,* 227.

Shi, J., Kantoff, P. W., Wooster, R., & Farokhzad, O. C. (2017). Cancer nanomedicine: Progress, challenges and opportunities. *Nature Reviews Cancer, 17,* 20–37.

Shioyama, H. (2001). Cleavage of graphite to graphene. *Journal of Materials Science Letters, 20,* 499–500.

Smalley, R. E. (1997). Discovering the fullerenes (Nobel lecture). *Angewandte Chemie, International Edition in English, 36,* 1594–1601.

Smith, A. D. (2013). Big moment for nanotech: Oncology therapeutics poised for a leap. *OncLive.*

Stankevich, I. V., Nikerov, M. V. E., & Bochvar, D. A. E. (1984). The structural chemistry of crystalline carbon: Geometry, stability, and electronic spectrum. *Russian Chemical Reviews, 53,* 640.

Stathopoulos, G. P., Antoniou, D., Dimitroulis, J., Michalopoulou, P., Bastas, A., Marosis, K., et al. (2010). Liposomal cisplatin combined with paclitaxel versus cisplatin and paclitaxel in non-small-cell lung cancer: A randomized phase III multicenter trial. *Annals of Oncology, 21,* 2227–2232.

Stavrovskaya, A. A. (2000). Cellular mechanisms of multidrug resistance of tumor cells. *Biochemistry (Mosc), 65,* 95–106.

Van Bommel, A., Crombeen, J., & Van Tooren, A. (1975). LEED and Auger electron observations of the SiC (0001) surface. *Surface Science, 48,* 463–472.

Venkatakrishnan, K., Liu, Y., Noe, D., Mertz, J., Bargfrede, M., Marbury, T., et al. (2014). Pharmacokinetics and pharmacodynamics of liposomal mifamurtide in adult volunteers with mild or moderate hepatic impairment. *British Journal of Clinical Pharmacology, 77,* 998–1010.

Viculis, L. M., Mack, J. J., & Kaner, R. B. (2003). A chemical route to carbon nanoscrolls. *Science, 299,* 1361.

Wicki, A., Witzigmann, D., Balasubramanian, V., & Huwyler, J. (2015). Nanomedicine in cancer therapy: Challenges, opportunities, and clinical applications. *Journal of Controlled Release, 200,* 138–157.

Chapter 2
Graphene-Based Nanomaterials: Introduction, Structure, Synthesis, Characterization, and Properties

2.1 Graphene-Based Nanomaterials

It is paramount to precisely define graphene-based nanomaterials (GBNs) to understand their structure-activity relationships and hence their applications, safety-hazards, and implications to human health. There are numerous carbon–based nanomaterials. All of them are not included in the GBNs. The graphene term has been used by scientists in a generic manner for graphene-based materials (Bianco et al. 2013). The GBNs include ultrafine graphite sheets (more than 10 graphene layers but, thickness should be less than 100 nm), graphene nanosheets (GNS), few-layer graphene (FLG), multi-layer graphitic nanosheets (MLG), graphite oxide, graphene oxide (GO), reduced graphene oxide (rGO), graphene nanoribbons (GNRs), and graphene quantum dots (GQDs).

The distinction among various GBNs can be made on the basis of morphological parameters (number of layers and average lateral size) and carbon-to-oxygen (C/O) atomic ratio (Fig. 2.1) (Wick et al. 2014). Number of layers of graphene attributes to the thickness of the material; therefore, it should be mentioned clearly wherever possible. Thickness and lateral dimension of the GBNs is a crucial parameter that affects the interactions of the GBNs with biological environment and events such as renal clearance, crossing the blood-brain barrier, cellular internalization, etc. (Sanchez et al. 2012; Russier et al. 2013). On the other hand, C/O ratio determines the surface functionalization and affects the dispersibility of the GBNs in variety of the solvents, biocompatibility, and colloidal behavior (Sanchez et al. 2012; Bianco 2013). These defining parameters can be measured by using the following analytical tools (see Table 2.1).

R. Srivastava et al., *Next Generation Graphene Nanomaterials for Cancer Theranostic Applications*,
https://doi.org/10.1007/978-981-33-6303-8_2

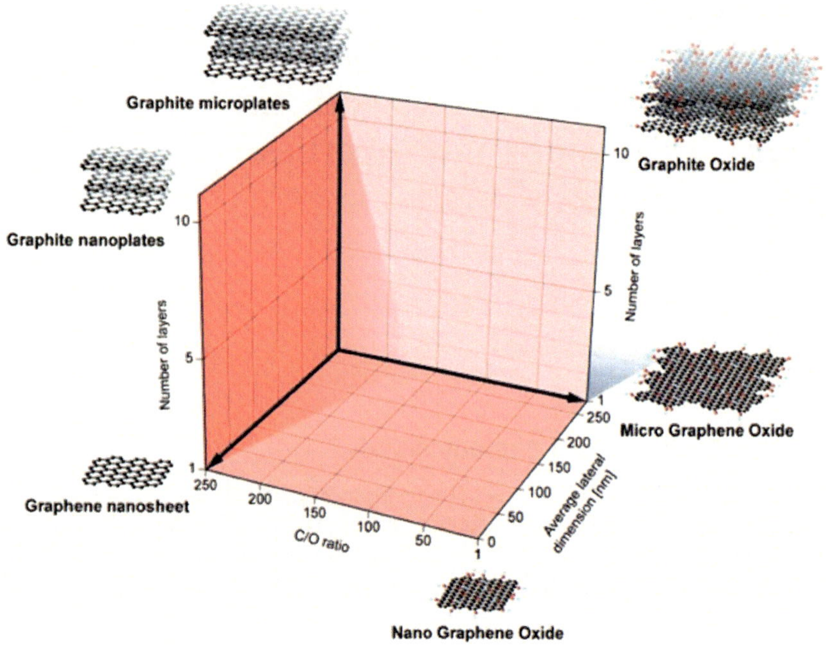

Fig. 2.1 An illustration is showing three important parameters of graphene derivatives to be discussed—lateral size, thickness, and functionalization. Adapted with permission from: Wick et al. (2014)

Table 2.1 Analytical tools that are used in analyzing the defining parameters for GBNs. Adapted with permission from: Wick et al. (2014)

Properties	Analytical technique
Lateral size	TEM, SEM, AFM
Number of layers (thickness)	TEM, AFM, Raman spectroscopy, optical absorbance measurements
C/O ratio	XPS, elemental analysis (ICP-MS)

2.1.1 Graphene

Graphene is a two-dimensional carbon nanomaterial with single-atomic thickness which acts as a building block for other GBNs. Carbon atoms in the graphene are sp^2 hybridized and connected in a honeycomb pattern. The discovery of graphene has been described in detail in the previous chapter. Graphene looks like a transparent sheet under FEG-TEM and atoms are seen arranged in a hexagonal lattice pattern at high-resolution observation (Fig. 2.2) (Geim and Novoselov 2007). FEG-TEM does not differentiate between monolayer graphene and a stack of two or more layers. For that, AFM and Raman spectroscopy tools are helpful.

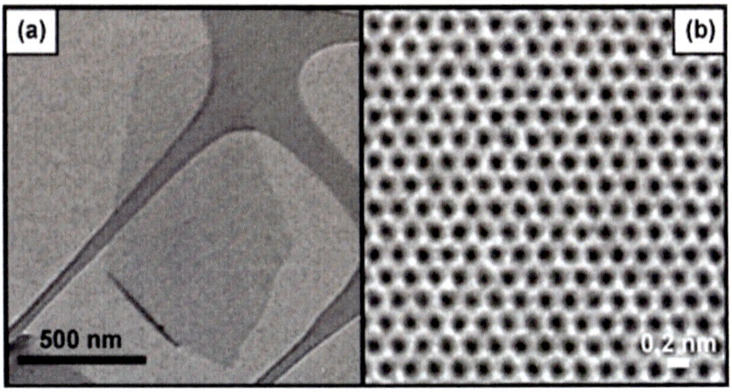

Fig. 2.2 FEG-TEM images of graphene **a** A transparent graphene sheet at low resolution **b** HR-TEM image that shows atoms arranged in a hexagonal lattice pattern. Adapted with permission from: Geim and Novoselov (2007)

In Raman spectroscopy, graphene shows three characteristic bands: D band (defect band) at ~1350 cm^{-1}, G band (graphitic band) at 1580 cm^{-1}, and 2D band of G′ band (overtone band) at 2700 cm^{-1}(Fig. 2.3a) (Georgakilas et al. 2015). The G band is attributed to the E_{2g} phonon of sp^2 hybridized carbon atoms whereas D band is credited for A_{1g} mode breathing vibrations of aromatic rings that are directly connected with sp^3 hybridized carbon atoms. The presence of sp^3 hybridized carbon atoms in the plane causes defects and that defines the structural purity of graphene. Therefore, the presence of D band in the spectrum depends on the quality of the graphene (Fig. 2.3a) (Ferrari and Basko 2013). The quality of graphene depends on its synthesis method. Graphene produced by liquid exfoliation of graphite will show no or least defects than the graphene obtained from reduction of graphene oxide. The 2D band appears due to overtone of D band and it signifies the presence of number of graphene monolayers. In case of graphite, it looks broad and divided into two peaks with intensity ratio of 1:2 appearing at around 2700 cm^{-1}. A pure graphene (single layer) Raman spectrum shows a single sharp 2D band before 2700 cm^{-1}. However, as the number of monolayers increases, 2D band broadens and appears at higher wavenumbers (Fig. 2.3b–c) (Ferrari et al. 2006). Graphene monolayers are separated from the ultrasonic liquid exfoliation of the graphite in selective organic solvents and fine dispersions can be achieved in N-methyl-2-pyrrolidone (NMP), pyridine, o-dichlorobenzene, and dimethylformamide (Bourlinos et al. 2009; De et al. 2010; Georgakilas et al. 2013; Hamilton et al. 2009; Khan et al. 2010). In another liquid synthesis route, graphene nanosheets are synthesized by reducing the GO using various reducing agents, hydrazine hydrate is most common. Other methods of graphene production include micromechanical, CVD, thermal annealing, and epitaxial growth (Georgakilas et al. 2015) (Fig. 2.4).

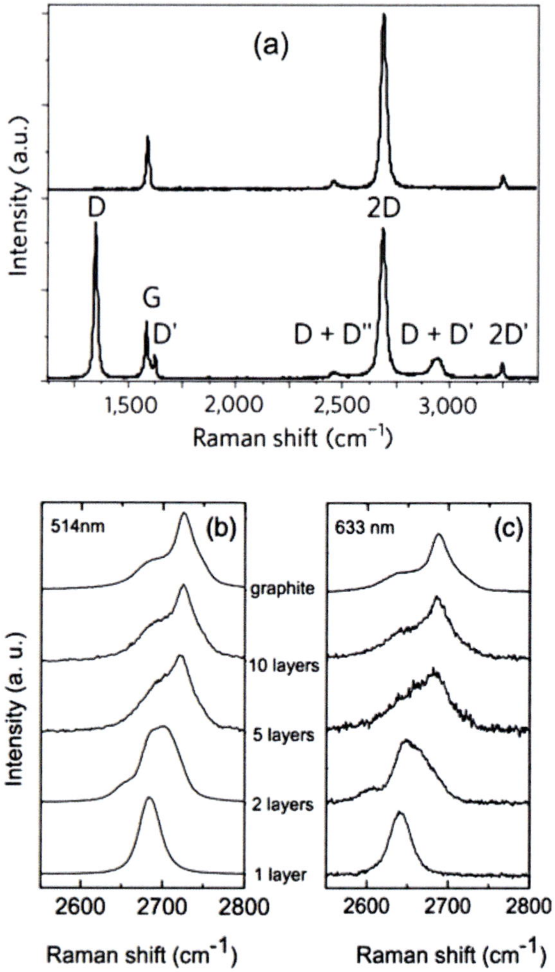

Fig. 2.3 Raman spectra of graphene: **a** pristine and defected (bottom) graphene. Adapted with permission from: Ferrari and Basko (2013) **b**, **c** evolution of the Raman spectra at 514 and 633 nm, respectively, with number of layers. Adapted with permission from: Ferrari et al. (2006)

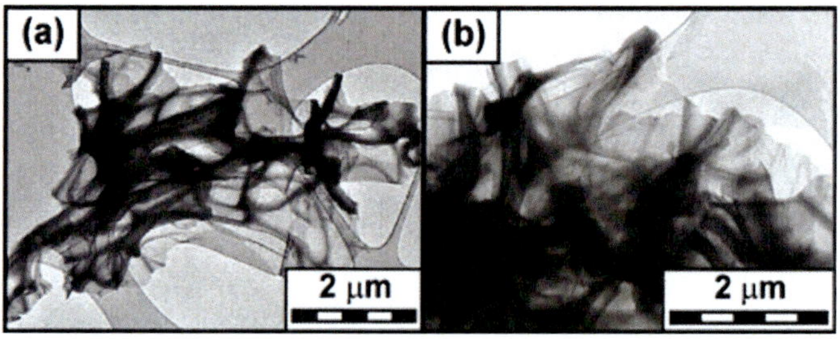

Fig. 2.4 FEG-TEM images of multi-layer graphitic nanosheets. Adapted with permission from: Bourlinos et al. (2009b)

2.1.2 Multi-layer Graphitic Nanosheets (MLGs)

MLGs are a stack of 2–10 graphene monolayers that have more or less similar to graphene in their properties. These can be identified by FEG-TEM and Raman spectroscopy. In FEG-TEM, they look like a dense stack of graphene monolayers. In Raman spectroscopy, as shown in Fig. 2.3, exhibits a shift in their 2D band and an obvious D band indicating multiple layers and defective sheets. MLGs are also produced while the graphene production through liquid exfoliation of graphite in wide variety of solvents and synthesis via chemical vapor deposition technique (CVD). Therefore, MLGs are often considered as a by-product of graphene synthesis.

2.1.3 Graphene Nanoribbons (GNRs)

Graphene nanoribbons are strip-like carbon nanostructures that have finite size and distinct edges. The structure of edge termination of GNRs categorizes them into three types: (i) zigzag, (ii) armchair, and (iii) chiral GNRs (Fig. 2.5). GNRs are composed of sp^2 hybridized carbon atoms and can be prepared by unzipping of the CNTs. The unzipping of the MWCNTs is usually carried out in the presence of strong oxidizing conditions using sulfuric acid (H_2SO_4) and potassium permanganate ($KMnO_4$). While oxidation, the layers of MWCNTs open up in sequence and oxygen-containing functional groups such as carbonyl, hydroxyl, and carboxylic are introduced in the plane and at edges. Other strategies for preparing GNRs have been shown in Fig. 2.6 that include intercalation of MWCNTs followed by unzipping, catalytic cutting of MWCNTs in the presence of metallic NPs, electric cutting, and plasma cutting (Terrones et al. 2010).

2.1.4 Graphene Oxide

Graphene oxide (GO) is an oxidized derivative of graphene which has various oxygen-containing functional groups decorated over its surface, in the plane, and at the edges. Graphene's sp^2 hybridized carbon atoms convert into sp^3 hybridized carbon atoms upon addition of functional groups that introduces the defects and hydrophilic nature into the sheets (Kim et al. 2012). GO is the monolayer material whether graphite oxide is the bulk oxidized form of the graphite. Synthesis of graphite oxide was first done by Brodie in 1857 (Brodie 1859). He treated graphite slurry with an oxidizing mixture containing potassium chlorate ($KClO_3$) and nitric acid (HNO_3). Further, Staudenmaier proposed a modification with addition of H_2SO_4 (Staudenmaier 1898). The most popular method for GO synthesis was given by Hummers in 1958 which made use of $KMnO_4$, sodium nitrate ($NaNO_3$), and

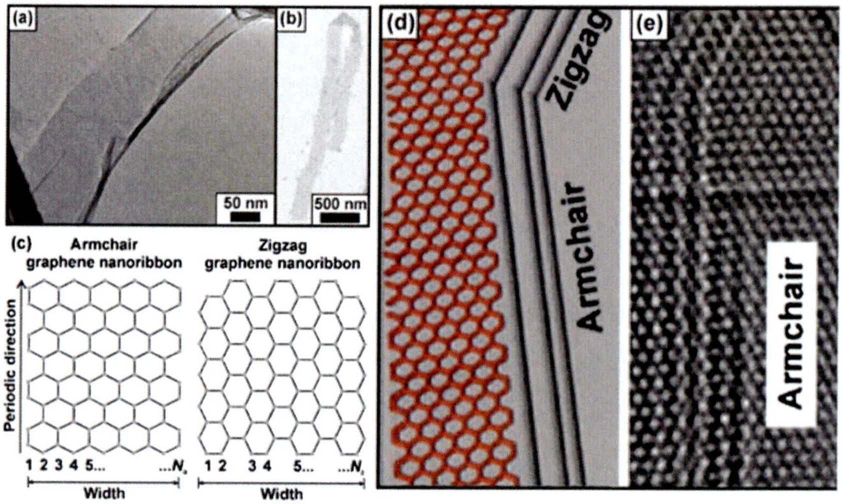

Fig. 2.5 GNRs: **a**, **b** FEG-TEM images, **c** structure illustration of armchair and zigzag GNRs, **d** model of a GNR edge showing a junction of armchair and zigzag edges, **e** HR-TEM image of GNR. Adapted with permission from: Georgakilas et al. (2015)

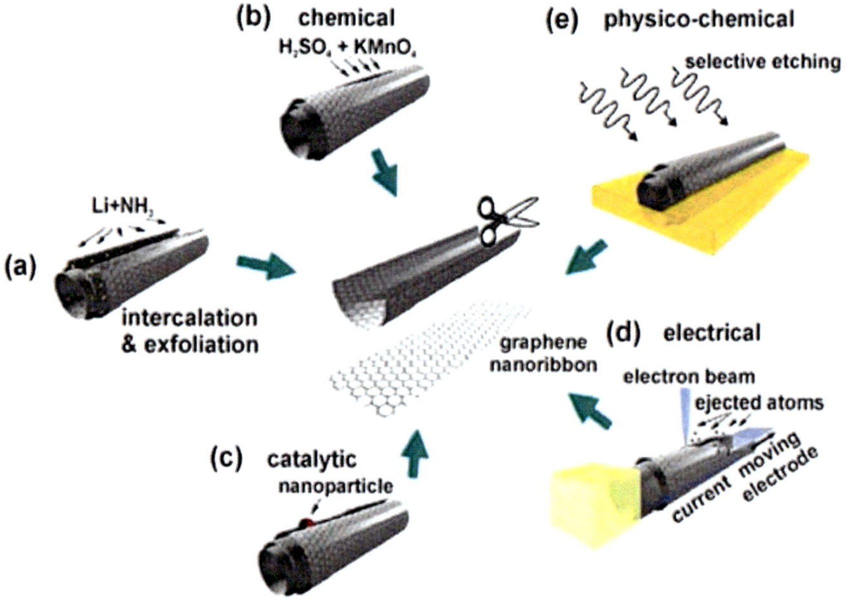

Fig. 2.6 Different strategies of unzipping of MWCNTs to produce GNRs. Adapted with permission from: Terrones et al. (2010)

H_2SO_4 to begin with the oxidation of the graphite (Hummers Jr and Offeman 1958). Thereafter, to date, there have been several modifications given by the scientists to improve the synthesis method (Chen et al. 2015, 2013, 2009; Marcano et al. 2010). The main focus of the modification in the GO synthesis method was to improve the yield of the product and to avoid the release of the toxic gases such as NO_2, N_2O_4, and ClO_2 during the reaction. Marcano and co-workers have done a magnificent contribution in this direction by suggesting the use of H_2SO_4 and H_3PO_4 in 9:1 ratio and $KMnO_4$ in the sixfold to the amount of graphite taken in the beginning (Fig. 2.7) (Marcano et al. 2010). The proposed modification improved the yield of GO and reduced the unreacted residue to a great extent.

After exfoliation and oxidation of graphite, functional groups such as carbonyl, carboxylic, epoxy, and hydroxyl groups are introduced that changes the property of the material. A few structural models for GO were proposed as shown in Fig. 2.8 (Gao et al. 2009; He et al. 1998; Lerf et al. 1998). Besides oxygen-containing functional groups, GO has hydrophobic patches too that imparts GO amphiphilic character. Thus, the GO is dispersible in wide variety of organic and aqueous solvents (Fig. 2.9) (Paredes et al. 2008). Physicochemical characterization of the GO is done using various techniques such as spectroscopy (UV-vis, FTIR, XPS, Raman, etc.), electron microscopy (FEG-SEM and FEG-TEM), and atomic force microscopy (AFM) (Kumawat et al. 2019). AFM microscopy is used to measure the dimensions and roughness of the GO sheets (Fig. 2.10a). In UV-vis spectroscopy, GO shows two signature peaks, one sharp peak at 232 nm, and one shoulder at 300 nm. FTIR of GO shows characteristic peaks due to functional

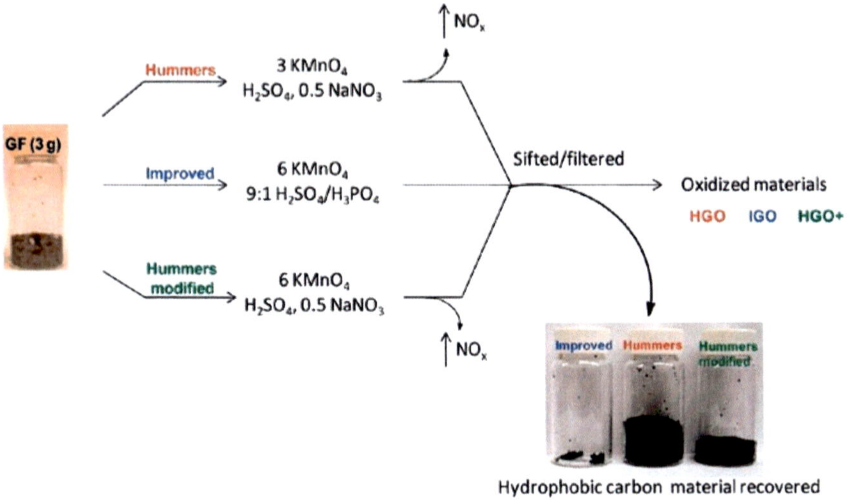

Fig. 2.7 Modification in Hummers' method for GO synthesis proposed by Marcano and co-workers. Adapted with permission from: Marcano et al. (2010)

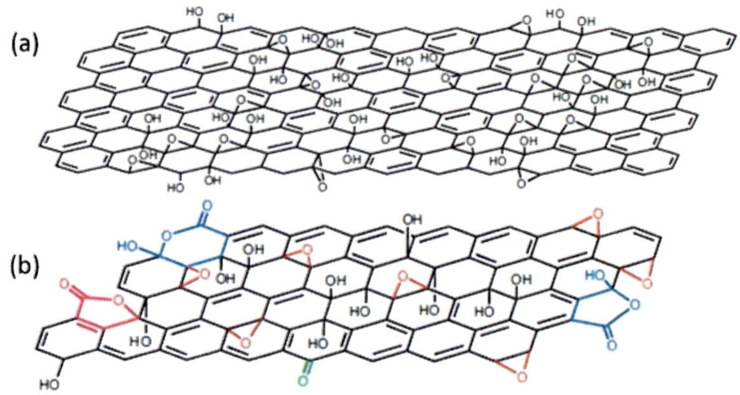

Fig. 2.8 Structure of graphene oxide: **a** Lerf–Klinowski model. Adapted with permission from: He et al. (1998), **b** Structural model proposed by Gao and co-workers. Adapted with permission from: Gao et al. (2009)

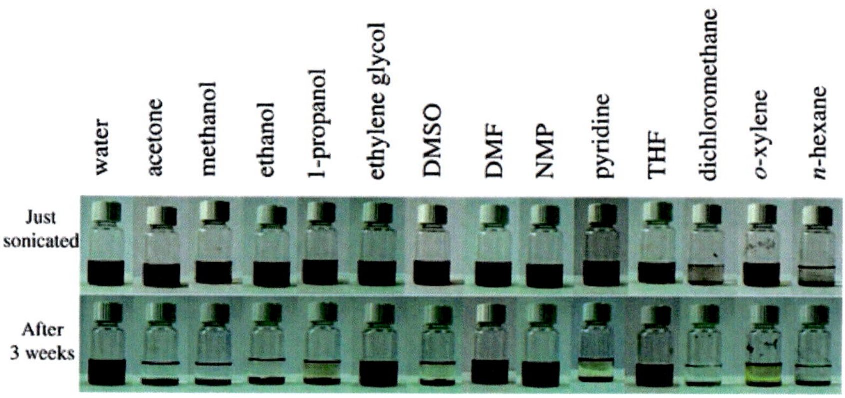

Fig. 2.9 Dispersions of GO in wide variety of solvents. Some of them are showing good stability. Adapted with permission from: Paredes et al. (2008)

groups stretching and bending vibrations (Fig. 2.10e). In Raman spectroscopy, GO shows two prominent bands, G and D bands at 1597 and 1355 cm^{-1}, respectively. As previously explained in Sect. 2.1.1, D band stands for the defects in the basal plane of the GO sheets due to oxidation of the graphite. Elemental analysis of GO can be done using XPS and ICP-MS. XPS gives the details about C/O ratio that tells about the degree of oxidation (Krishnamoorthy et al. 2013).

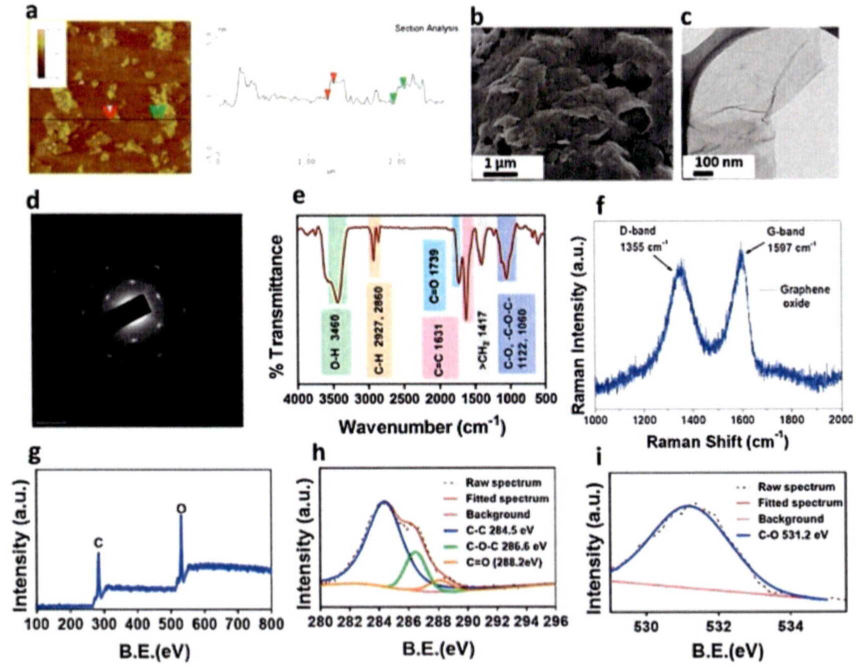

Fig. 2.10 Physicochemical characterization of GO using various techniques **a** AFM image **b** FEG-SEM image **c** FEG-TEM image **d** SAED pattern of GO **e** FTIR spectrum **f** Raman spectrum **g** XPS survey spectrum **h** XPS deconvoluted C 1 s spectrum **i** XPS deconvoluted O 1 s spectrum. Adapted with permission from: Kumawat et al. (2019)

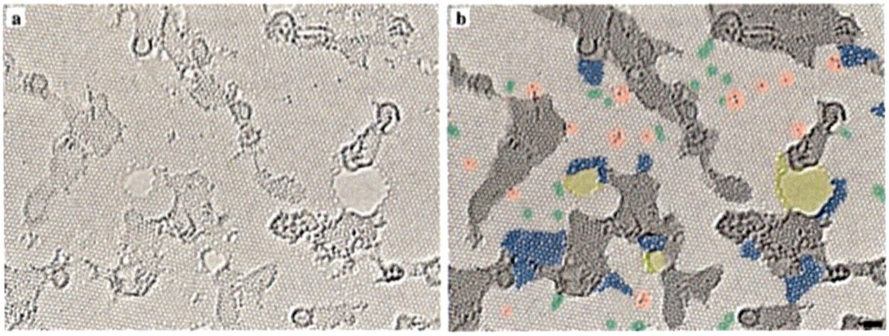

Fig. 2.11 FEG-TEM images of RGO: **a** original image **b** image shown with colors highlighting specific features of the sheet structure. Adapted with permission from: Gomez-Navarro et al. (2010)

2.1.5 Reduced Graphene Oxide (RGO)

RGO is achieved by reducing the oxygen-containing functional groups of the GO. The reduction of GO is considered an alternate route to achieve graphene. However, unlike graphene, RGO does contain topological defects due to elimination of the functional groups from the GO sheets. Such defects are not visible in the mechanically exfoliated graphene sheets. Figure 2.11 shows the HR-TEM images of RGO sheet, image 2.11b is with the features highlighted (Gomez-Navarro et al. 2010). The light gray area in Fig. 2.11 depicts a well-crystallized area akin to graphene and it produces a well-organized hexagonal SAED pattern. The areas shown with yellow highlights depict the holes and they are present in the mechanically exfoliated graphene too. However, the green patches are the isolated topological defects. The most common and widely used reducing agent for GO reduction is hydrazine hydrate (Stankovich et al. 2007). However, the high temperature conditions required for the reduction and toxic nature of the hydrazine hydrate are the shortcomings of this method. Therefore, researchers have explored various reduction strategies for the GO reduction. An account of various strategies has been tabulated in Table 2.2.

As the reduction of the GO progresses, the change in the color of the reaction mixture occurs from yellow-brown to dark brown and turns black with reaction time. In UV-vis spectrum, the peak which GO shows at 232 nm shifts to higher wavelengths toward 265 nm and the shoulder peak that appears at 300 nm shifts toward 325 nm (Fig. 2.12b). In XRD, the peak in GO at 10° gets disappeared after reduction (Fig. 2.12c). Thermograms are another data sets that help in investigating the degree of reduction by checking the weight loss of the material as function with increasing temperature. As shown in Fig. 2.13a, GO rapidly loses its weight as the temperature crosses 100 °C due to elimination of the water molecules and volatile oxygen functional groups (Moon et al. 2010). Conversely, RGO shows the thermal stability because of absence of any volatile functional groups. In Raman spectroscopy, as shown in Fig. 2.13b I_D/I_G ratio increases. After reduction, new domains of the sp^2-hybridized carbon atoms form and number of defects increases because of the elimination of functional groups. In XPS spectroscopy, the peaks related to C–C bond (near 284 eV) get intense and peaks attributed to C–O, C–O–C, >C=O, and –(O)C–O get minimized in C 1 s spectra (Fig. 2.13c–d).

2.1.6 Graphene Quantum Dots

Graphene quantum dots (GQDs) are fine-sized fluorescent nanomaterial that are made up of a few-layer graphene sheets and found commonly with lateral dimension mostly <10 nm, sometimes 10–20 nm, and rarely ~60 nm (Liu et al. 2011; Shen et al. 2012b). The GQDs are different from other two carbon-based nanodots (carbon dots and polymeric dots). GQDs are graphitic structures with

Table 2.2 Reduction strategies for GO

Reduction method/ reducing agent	Reaction conditions (temperature, time, pH, etc.)	Advantage/ disadvantage	References
Hydrazine hydrate	100 °C, 24 h	Efficient/toxic reagent	Stankovich et al. (2007)
NH_3/Temp.	300 °C, 25 min	Nitrogen-doped RGO/high temperature	Li et al. (2009)
Bovine serum albumin (BSA)	RT, 24 h	Room temperature, eco-friendly	Liu et al. (2010)
Hydroxyl amine ($NH_2OH.HCl$)	90 °C, 30 min	Rapid	Zhou et al. (2011)
Hydriodic acid with acetic acid (HI-AcOH)	40 °C, 40 h	Low temperature	Moon et al. (2010)
NaOH, KOH	50–90 °C	Green route	Fan et al. (2010)
Electrochemical	2 h	Green synthesis/ defects in the sheets	Guo et al. (2009)
Pulsed laser	RT, 5 min	Rapid/expensive setup	Liu et al. (2012)
Nascent hydrogen (Al foil/HCl, Al foil/ NaOH, Zn/NaOH)	RT, 20 min–6 h	No toxic reagents	Pham et al. (2012)
$NaBH_4$	24–48 h	The high conductivity of graphene/toxic reagent	Shin et al. (2009)
N, N-dimethylacetamide (DMAc)/microwave	165 °C, 10 min	Rapid	Chen et al. (2010)
Aluminum powder/ HCl	RT, 30 min	Environmental friendly	Fan et al. (2010)
Bacteria	RT, 24 h	Room temperature	Salas et al. (2010)
Alcohols	100 °C, 5 d	No toxic reagents/ time consuming	Dreyer et al. (2011)
Fe powder/HCl	RT, 6 h	No toxic reagents	Fan et al. (2011)
Melatonin	40–80 °C, 3 h, 9–10	Rapid, eco-friendly	Esfandiar et al. (2011)
Tea solution (polyphenols)	100 °C	Reduction cum stabilization	Wang et al. (2011)
UV irradiation	24–48 h	Green approach	Ding et al. (2011)
Pyrrole	95 °C, 12 h	Reduction cum dispersion	Amarnath et al. (2011)
Baker's yeast/NADPH	35–40 °C, 72 h	Eco-friendly approach	Khanra et al. (2012)

(continued)

Table 2.2 (continued)

Reduction method/ reducing agent	Reaction conditions (temperature, time, pH, etc.)	Advantage/ disadvantage	References
Carrot root	25 °C, 48 h	Eco-friendly approach	Kuila et al. (2012)
Urea	95 °C	No toxic reagents	Lei et al. (2012)
Non-aromatic amino acids	90 °C, 12–13 h	No toxic reagents	Tran et al. (2014)
Neem leaves extract	100 °C, 72 h	Eco-friendly	Babu et al. (2014)
IR-irradiation		No toxic reagents	Guo et al. (2013)
Na-NH_3	1 h	Efficient, low temperature/ sodium metal	Feng et al. (2013)
Supercritical alcohols	400 °C, 2–3 h	No toxic reagents	Seo et al. (2013)
Malt	90 °C, 5 h	Eco-friendly	Tabrizi et al. (2014)
Vegetable (ladies finger) extract	90 °C, 12 h	Eco-friendly	Gnanaprakasam and Selvaraju (2014)
Temperature	400 °C, 1 h, vacuum	Efficient and green route	Kim et al. (2015)
Caffeic acid	95 °C, 24 h	Green route	Bo et al. (2014)
Gamma rays		Green and dry route	Dumée et al. (2014)
Ascorbic acid	95 °C, 1 h	Rapid and scalable	Abdolhosseinzadeh et al. (2015)
Temperature, microwave	300 °C, 1 h, 1–2 s	Rapid, highly efficient	Voiry et al. (2016)
Bacteriorhodopsin	39 °C, 3 h	Green route	Akhavan (2015)
UV irradiation/Sulfite photolysis	1 h	Environment friendly, cost effective	Yin et al. (2019)
In situ reduction/KOH	–	Nontoxic approach	Tang et al. (2019)
Kaffir lime peel extraction	8–10 h	Green reduction	Wijaya et al. (2020)

crystalline property wherein most of the carbon atoms are in sp^2-hybridization state (Fig. 2.14d–e). Thickness of the GQDs is found to be 1–3 graphene layers (Fig. 2.14). The GQDs are synthesized by following two approaches: (a) top-down approach and (b) bottom-up approach (Fig. 2.15). In top-down approach, bulk precursor is cut down into fine-sized particles to form the GQDs, whereas in bottom-up approach, small molecules are covalently add to each other and form the GQDs. Top-down approach mainly includes methods such as chemical cutting of graphene derivatives, electrochemical synthesis, and oxygen plasma treatment. However, bottom-up approach includes such as fullerene cage-opening, solution

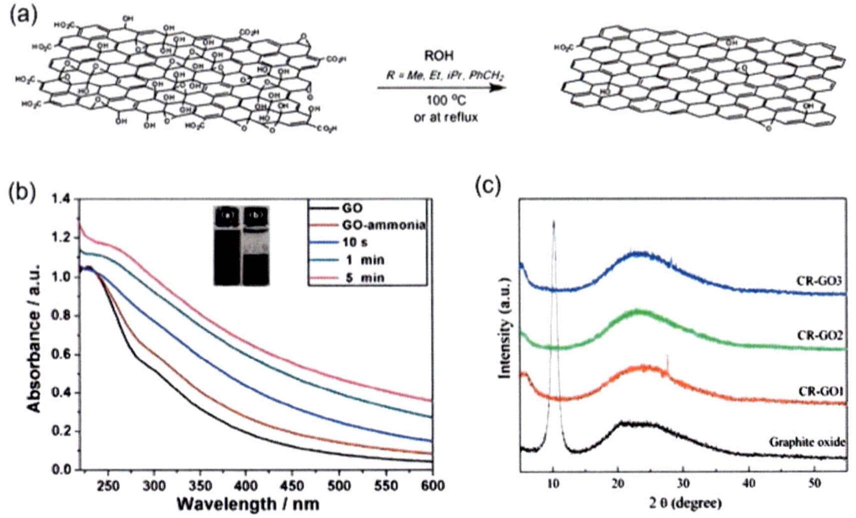

Fig. 2.12 **a** A reduction strategy of GO using alcohols, the structure of RGO depicting the absence of functional groups after reduction. Adapted with permission from: Dreyer et al. (2011) **b** comparative UV-vis spectra. Adapted with permission from: Huang et al. (2011) **c** comparative XRD spectra. Adapted with permission from: Kuila et al. (2012)

chemistry-based methods, and fabrication of the GQDs from carbon atomic beam (Shen et al. 2012b). First report of the GQDs synthesis came in 2010 wherein Pan and co-workers reported the synthesis of the GQDs from hydrothermal cutting of the graphene sheets at 200 °C for 10 h in the presence of NaOH (pH = 8.0) (Fig. 2.16a) (Pan et al. 2010). It was a top-down approach. During hydrothermal treatment of GO, the chains of epoxy groups convert into epoxy pairs that further convert into energetically favorable carbonyl functional groups. The presence of functional groups in such fashion creates linear defects in the sheets and makes them vulnerable toward deoxidization process. Further, deoxidization process breaks the sheets into fragments and produce the GQDs (Fig. 2.16a) (Pan et al. 2010). In hydrothermal cleaving of the sheets, though the position of sites of the cleavage is not at certain interval, therefore, it produces the GQDs with varying sizes. But, when the cleaving of the sheets is carried out using chemical scissors such as PEG, the resulting GQDs become nearly monodisperse. In this case, PEG acts as a chemical scissor and a stabilizing agent as well (Fig. 2.16c–d) (Xue et al. 2013). In the bottom-up methods, carbonization or pyrolysis of small molecules (precursors) occurs under heat treatment. The heating is carried out through combustion or plasma or microwave treatment. The precursors for the GQDs synthesis via bottom-up approach include majorly organic compounds such as glutathione, ascorbic acid, glutamic acid, ethylenediaminetetraacetic acid disodium salt (EDTA-2Na), citric acid, pyrene derivatives, and plant extracts (Dong et al. 2012; Ju and Chen 2014; Kumawat et al. 2017b, 2017a; Zheng et al. 2015). Various

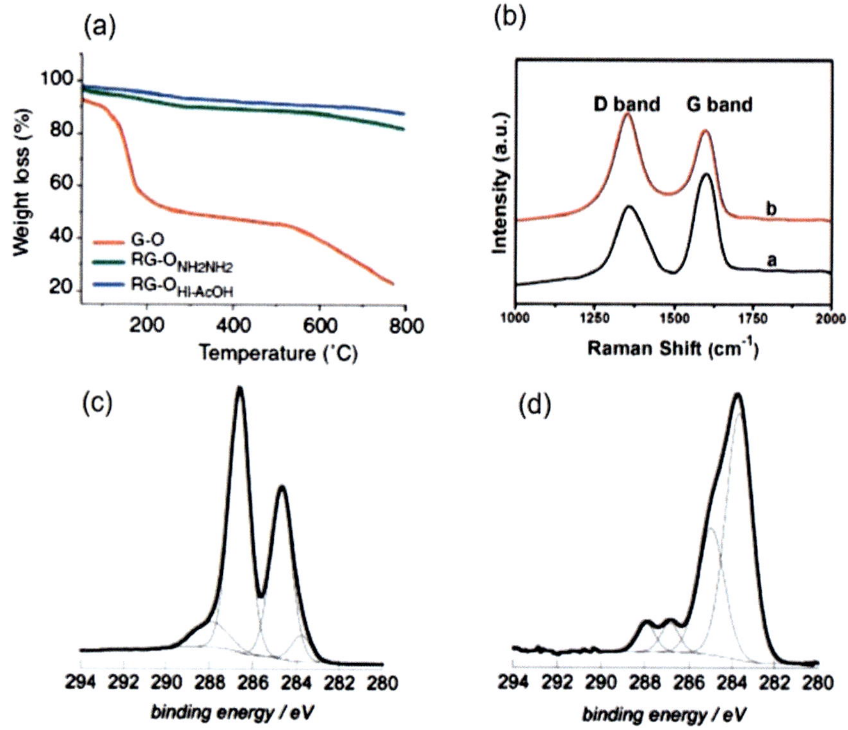

Fig. 2.13 **a** Comparative TGA thermograms. Adapted with permission from: Moon et al. (2010), **b** comparative Raman spectra (red line-RGO, black line-GO). Adapted with permission from: Zhou et al. (2011), **c** Deconvoluted XPS C 1 s spectra of GO, and **d** deconvoluted XPS C 1 s spectra of RGO. Adapted with permission from: Kaminska et al. (2012)

strategies of preparing the GQDs have been summarized in Table 2.3. For purification of the GQDs, methods such as column chromatography, gel-electrophoresis, dialysis, ultra-filtration, and centrifugation are adopted (Li et al. 2010; Liu et al. 2007; Zheng et al. 2015).

Due to quantum confinement and edge effects, the GQDs have nonzero band gaps and possess amazing electronic and optical properties. The GQDs exhibit tuneable fluorescence property that is governed by their size, shape, geometry, surface passivation, degree of oxidation, and their edge structure (Georgakilas et al. 2015). These factors are broadly classified into two categories: (a) intrinsic state emission (b) defect state emission. When the oxidative cleavage of the graphene occurs, it leaves two kinds of the edge structures behind: (a) arm-chair and (b) zigzag (Fig. 2.17a). However, the chemically derived graphene sheets have both free and functionalized edges in coexistence. The zigzag sites, as shown in Fig. 2.17 (right) edge structures proposed by Radovic and Bockrath, are carbene-like with the triplet ground state most common and armchair sites are carbene-like with the singlet state being most common.

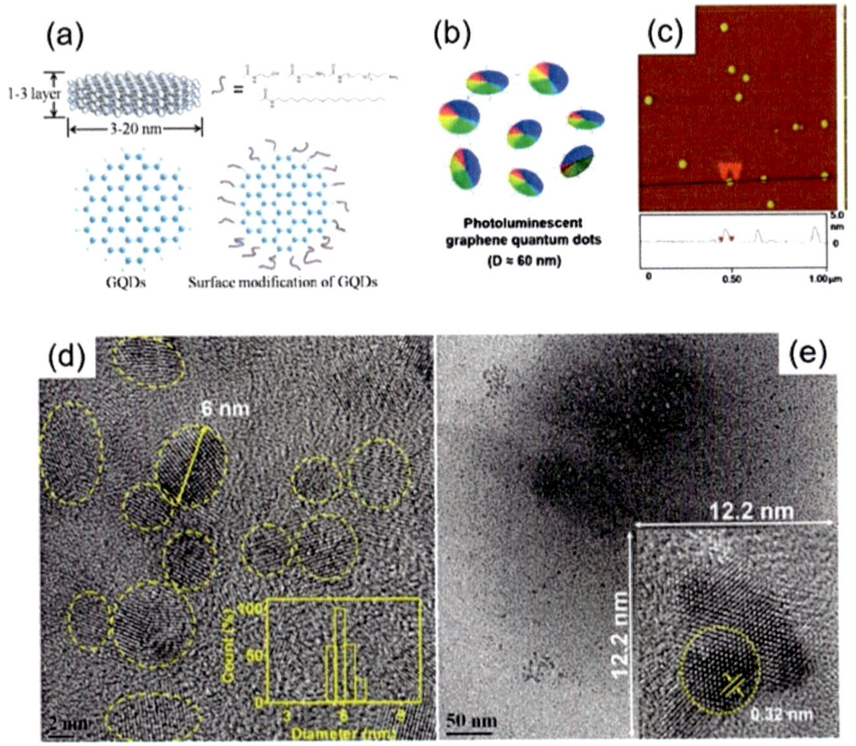

Fig. 2.14 Structure of the GQDs, surface modification, and FEG-TEM images. Adapted with permission from: Kumawat et al. (2017b); Liu et al. (2011)

Pan and co-workers suggested that the blue photoluminescence (PL) is originated due to zigzag sites of the GQDs. The carbene-like zigzag sites have the electronic configuration $\sigma^1\pi^1$ wherein, its highest occupied molecular orbitals (HOMO) are singly occupied, showing the triplet ground state and that was evident because of the peaks shown in the absorption spectrum at 320 nm (3.86 eV) and 257 nm (4.82 eV). They also suggested that these carbene-like sites get protonated under acidic environment and due to that PL gets quenched which is restored again in alkali medium (Fig. 2.16b) (Pan et al. 2010). The GQDs absorb in the UV and visible region and show peaks correspond to π–π* and n–π* transitions that are attributed to double bonds in the basal plane and heteroatom-containing functional groups. The GQDs show excitation-dependent as well as excitation-independent emissions (Kumawat et al. 2017a, 2017b; Zheng et al. 2015). Apart from the edge structure, the PL properties are also affected by the core of the GQDs. The GQDs which have lesser number of functional groups, their PL property is affected by their core structure that acts as PL center. Larger the size of the core, smaller the bag gap become, as shown in Fig. 2.18. In graphene, however, the conjugated π-domain is large that it leads to narrowing of the band gap. Mueller and co-workers reported

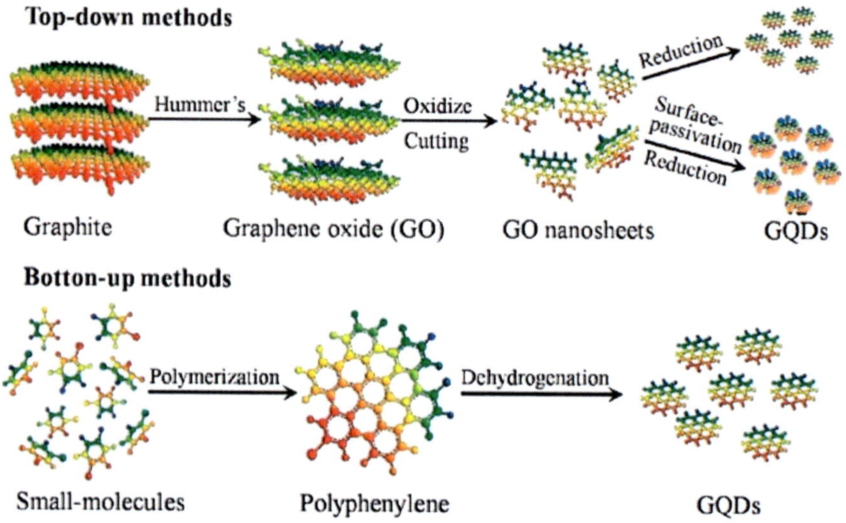

Fig. 2.15 Two approaches for synthesis of the GQDs-Top-down approach (top) and bottom-up approach (bottom). Adapted with permission from: Shen et al. (2012b)

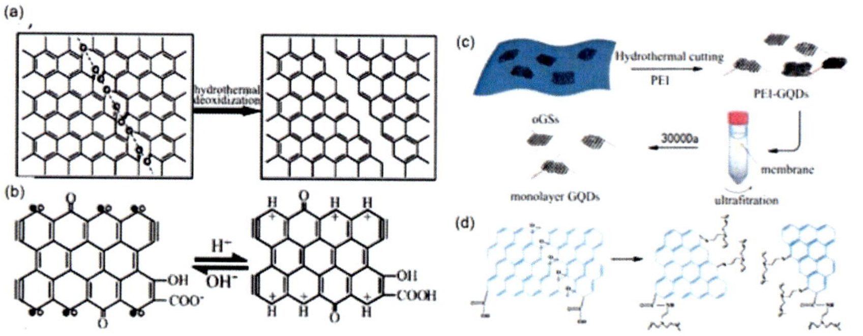

Fig. 2.16 Hydrothermal cutting of graphene nanomaterials to produce the GQDs. Adapted with permission from: Pan et al. (2010); Xue et al. (2013)

that the GQDs show excellent intersystem crossing (ISC) as well as it competes with internal conversion (IC) because of very narrow (ca. 175 meV) splitting among the singlet-triplet states (Mueller et al. 2010). Therefore, GQDs show fluorescence as well as phosphorescence.

The size of the GQDs affects optical properties of the GQDs more than their shapes; however, these two factors can be governed by the stepwise synthesis and annealing temperature. The presence of oxygen atomic functional groups acts as emissive traps that shrink the quantum yield (QY). However, reducing these functional groups or passivation of the surface of the GQDs using polymers such as

Table 2.3 Different strategies of preparation of GQDs

Synthesis approach	Type of synthesis method	Precursor or starting compound	Synthesis method/reaction conditions	GQDs size range (average diameter) nm	PL type/maxima or range	References
Top-down approach	From cutting of graphene and graphene derivatives	GO and Graphene sheets	Hydrothermal method, 200–300 °C, 2 h, nitrogen atmosphere/H_2SO_4 + HNO_3, 15–20 h, ultrasonication (500 W)/NaOH, pH = 8.0, 200 °C, 10 h	~9.6 nm	Blue, excitation-dependent/ 430 nm	Pan et al. (2010)
		GO and polyethylenegl-ycol (PEG)	Hydrothermal method, GO + HNO_3, 70 °C, 24 h/PEG, 200 °C, 24 h	5–25 nm	Blue excitation-dependent	Shen et al. (2012b)
		GO	Reduction reaction, GO + HNO3/ GO + PEG diamine/Hydrazine hydrate	5–19 nm (13.3)	Blue, excitation-dependent	Shen et al. (2011)
		GO	Hydrothermal method/GO + PEI, 150 °C	2–4 nm (3.3)	Blue, excitation-independent	Xue et al. (2013)
	Electrochemical synthesis	Graphene film as a working electrode	–	3–5 nm	Green luminescence	Li et al. (2011)
		Graphite	Graphite anode in alkaline solution, hydrazine hydrate	5–10 nm	Yellow, excitation-independent, 540 nm	Zhang et al. (2012)
		MWCNTs	MWCNTs in propylene carbonate, $LiClO_4$, 90 °C/30 °C	3–8.5 nm	Excitation-dependent	Shinde and Pillai (2012)
	Oxygen plasma treatment	Graphene	Graphene exposed to oxygen-argon (1:2) RF plasma		–	Gokus et al. (2009)

(continued)

Table 2.3 (continued)

Synthesis approach	Type of synthesis method	Precursor or starting compound	Synthesis method/reaction conditions	GQDs size range (average diameter) nm	PL type/maxima or range	References
Bottom-up approach	Fullerene cage-opening	C_{60} fullerenes	Ruthenium catalyzed cage-opening	–	–	Lu et al. (2011)
	Solution chemistry-based methods	hexa-peri-hexabenzocoronene (HBC)	Pyrolysis, 600–1200 °C,	~60 nm	Blue, excitation-dependent	Liu et al. (2011)
		Pyrene derivatives	Nitration of pyrene/hydrothermal treatment/ HNO_3, 80 °C/NaOH, 200 °C, 10 h	3–4 nm	Bright green, excitation-dependent	Wang et al. (2014)
		Mango-leaf extracts	Microwave-heat treatment	2–8 nm	Red, excitation-independent, 680 nm	Kumawat et al. (2017a, 2017b)
	From carbon atomic beam	Ethylene gas	Argon plasma treatment of ethylene gas	7–23 nm	Blue, excitation-dependent	Kim and Suh (2014)

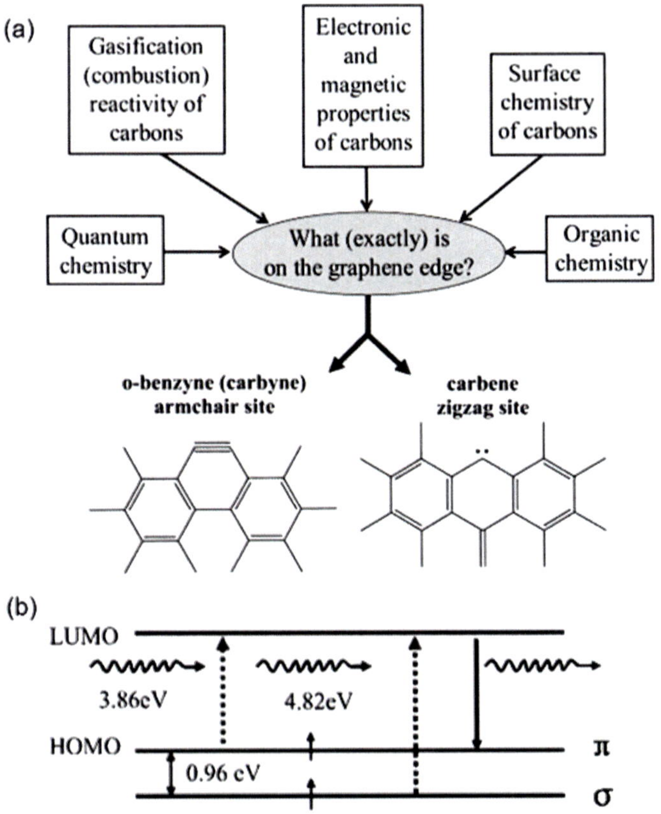

Fig. 2.17 **a** Edge structures of sp^2-hybridized carbon materials proposed by Radovic and Bockrath. Adapted with permission from: Radovic and Bockrath (2005), **b** electronic transitions from a zigzag site. Adapted with permission from: Pan et al. (2010)

PEG, polyethyleneimine (PEI), and branched polyethyleneimine (BPEI) enhances the QY (Li et al. 2012; Zhu et al. 2012). In other reports, it was shown that functionalization of the GQDs with amino group lowers the band gap (Fig. 2.19). It is because of establishment of the orbital resonance with π orbitals of the core. The increase in the number of the amino groups further lowers the band gap of the GQDs (Jin et al. 2013; Tetsuka et al. 2012).

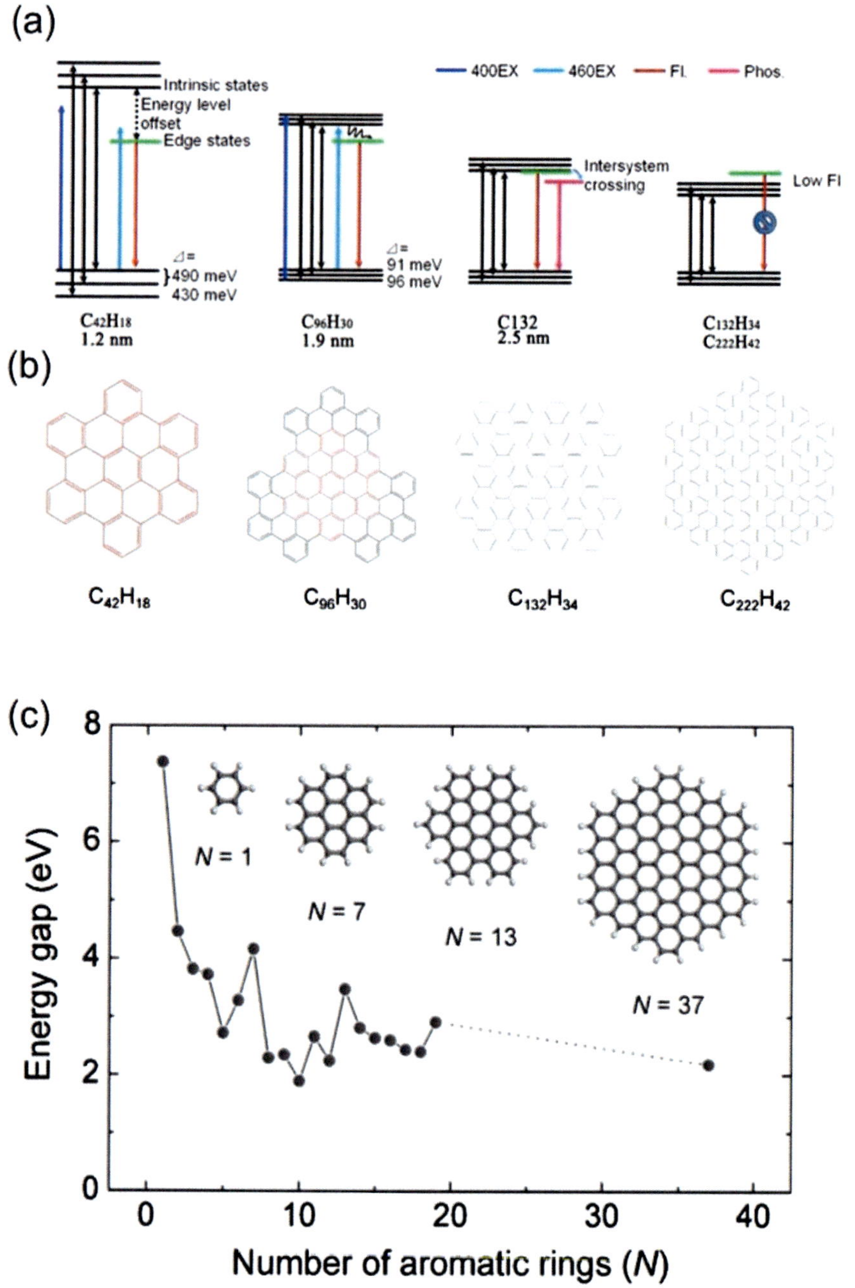

Fig. 2.18 Quantum confinement effect in the GQDs. Adapted with permission from: Eda et al. (2010); Zhu et al. (2014)

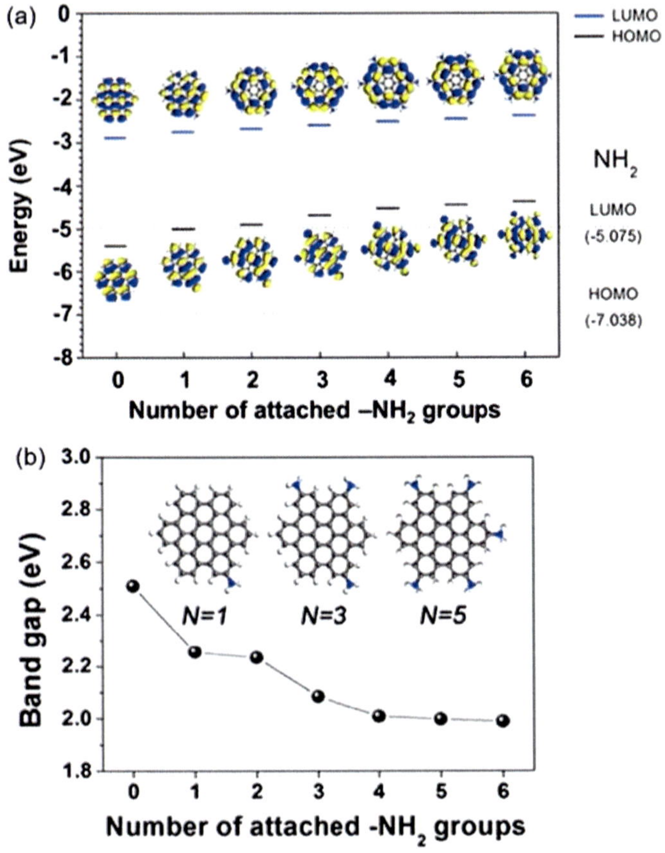

Fig. 2.19 Effect of amino functional groups on the band gap of the GQDs. Adapted with permission from: Jin et al. (2013); Tetsuka et al. (2012)

References

Abdolhosseinzadeh, S., Asgharzadeh, H., & Kim, H. S. (2015). Fast and fully-scalable synthesis of reduced graphene oxide. *Scientific Reports, 5,* 10160.

Akhavan, O. (2015). Bacteriorhodopsin as a superior substitute for hydrazine in chemical reduction of single-layer graphene oxide sheets. *Carbon, 81,* 158–166.

Amarnath, C. A., Hong, C. E., Kim, N. H., Ku, B.-C., Kuila, T., & Lee, J. H. (2011). Efficient synthesis of graphene sheets using pyrrole as a reducing agent. *Carbon, 49,* 3497–3502.

Babu, K. J., Nahm, K. S., & Hwang, Y. J. (2014). A facile one-pot green synthesis of reduced graphene oxide and its composites for non-enzymatic hydrogen peroxide sensor applications. *RSC Advances, 4,* 7944–7951.

Bianco, A. (2013). Graphene: Safe or toxic? The two faces of the medal. *Angewandte Chemie International Edition in English, 52,* 4986–4997.

Bianco, A., Cheng, H.-M., Enoki, T., Gogotsi, Y., Hurt, R. H., Koratkar, N., Kyotani, T., Monthioux, M., Park, C. R., & Tascon, J. M. (2013). All in the graphene family–A recommended nomenclature for two-dimensional carbon materials. Elsevier.

Bo, Z., Shuai, X., Mao, S., Yang, H., Qian, J., Chen, J., et al. (2014). Green preparation of reduced graphene oxide for sensing and energy storage applications. *Scientific Reports, 4,* 4684.

Bourlinos, A. B., Georgakilas, V., Zboril, R., Steriotis, T. A., & Stubos, A. K. (2009a). Liquid-phase exfoliation of graphite towards solubilized graphenes. *Small, 5,* 1841–1845.

Bourlinos, A. B., Steriotis, T. A., Zboril, R., Georgakilas, V., & Stubos, A. (2009b). Direct synthesis of carbon nanosheets by the solid-state pyrolysis of betaine. *Journal of materials Science, 44,* 1407–1411.

Brodie, B. C. (1859). XIII. On the atomic weight of graphite. *Philosophical Transactions of the Royal Society of London,* 249–259.

Chen, J., Li, Y., Huang, L., Li, C., & Shi, G. (2015). High-yield preparation of graphene oxide from small graphite flakes via an improved Hummers method with a simple purification process. *Carbon, 81,* 826–834.

Chen, J., Yao, B., Li, C., & Shi, G. (2013). An improved Hummers method for eco-friendly synthesis of graphene oxide. *Carbon, 64,* 225–229.

Chen, T., Zeng, B., Liu, J., Dong, J., Liu, X., Wu, Z., Yang, X., & Li, Z. (2009) High throughput exfoliation of graphene oxide from expanded graphite with assistance of strong oxidant in modified Hummers method. *Journal of Physics: Conference Series.* 012051.

Chen, W., Yan, L., & Bangal, P. R. (2010). Preparation of graphene by the rapid and mild thermal reduction of graphene oxide induced by microwaves. *Carbon, 48,* 1146–1152.

De, S., King, P. J., Lotya, M., O'Neill, A., Doherty, E. M., Hernandez, Y., et al. (2010). Flexible, transparent, conducting films of randomly stacked graphene from surfactant-stabilized, oxide-free graphene dispersions. *Small, 6,* 458–464.

Ding, Y. H., Zhang, P., Zhuo, Q., Ren, H. M., Yang, Z. M., & Jiang, Y. (2011). A green approach to the synthesis of reduced graphene oxide nanosheets under UV irradiation. *Nanotechnology, 22,* 215601.

Dong, Y., Shao, J., Chen, C., Li, H., Wang, R., Chi, Y., et al. (2012). Blue luminescent graphene quantum dots and graphene oxide prepared by tuning the carbonization degree of citric acid. *Carbon, 50,* 4738–4743.

Dreyer, D. R., Murali, S., Zhu, Y., Ruoff, R. S., & Bielawski, C. W. (2011). Reduction of graphite oxide using alcohols. *Journal of Materials Chemistry, 21,* 3443–3447.

Dumée, L. F., Feng, C., He, L., Yi, Z., She, F., Peng, Z., et al. (2014). Single step preparation of meso-porous and reduced graphene oxide by gamma-ray irradiation in gaseous phase. *Carbon, 70,* 313–318.

Eda, G., Lin, Y. Y., Mattevi, C., Yamaguchi, H., Chen, H. A., Chen, I. S., et al. (2010). Blue photoluminescence from chemically derived graphene oxide. *Advanced Materials, 22,* 505–509.

Esfandiar, A., Akhavan, O., & Irajizad, A. (2011). Melatonin as a powerful bio-antioxidant for reduction of graphene oxide. *Journal of Materials Chemistry, 21,* 10907–10914.

Fan, Z., Wang, K., Wei, T., Yan, J., Song, L., & Shao, B. (2010). An environmentally friendly and efficient route for the reduction of graphene oxide by aluminum powder. *Carbon, 48,* 1686–1689.

Fan, Z. J., Kai, W., Yan, J., Wei, T., Zhi, L. J., Feng, J., et al. (2011). Facile synthesis of graphene nanosheets via Fe reduction of exfoliated graphite oxide. *ACS Nano, 5,* 191–198.

Feng, H., Cheng, R., Zhao, X., Duan, X., & Li, J. (2013). A low-temperature method to produce highly reduced graphene oxide. *Nature Communications, 4,* 1539.

Ferrari, A. C., & Basko, D. M. (2013). Raman spectroscopy as a versatile tool for studying the properties of graphene. *Nature Nanotechnology, 8,* 235–246.

Ferrari, A. C., Meyer, J. C., Scardaci, V., Casiraghi, C., Lazzeri, M., Mauri, F., et al. (2006). Raman spectrum of graphene and graphene layers. *Physical Review Letters, 97,* 187401.

Gao, W., Alemany, L. B., Ci, L., & Ajayan, P. M. (2009). New insights into the structure and reduction of graphite oxide. *Nature Chemistry, 1,* 403–408.

Geim, A. K., & Novoselov, K. S. (2007). The rise of graphene. *Nature Materials, 6,* 183–191.

Georgakilas, V., Kouloumpis, A., Gournis, D., Bourlinos, A., Trapalis, C., & Zboril, R. (2013). Tuning the dispersibility of carbon nanostructures from organophilic to hydrophilic: Towards the preparation of new multipurpose carbon-based hybrids. *Chemistry, 19,* 12884–12891.

Georgakilas, V., Perman, J. A., Tucek, J., & Zboril, R. (2015). Broad family of carbon nanoallotropes: classification, chemistry, and applications of fullerenes, carbon dots, nanotubes, graphene, nanodiamonds, and combined superstructures. *Chemical Reviews, 115,* 4744–4822.

Gnanaprakasam, P., & Selvaraju, T. (2014). Green synthesis of self assembled silver nanowire decorated reduced graphene oxide for efficient nitroarene reduction. *RSC Advances, 4,* 24518–24525.

Gokus, T., Nair, R. R., Bonetti, A., Bohmler, M., Lombardo, A., Novoselov, K. S., et al. (2009). Making graphene luminescent by oxygen plasma treatment. *ACS Nano, 3,* 3963–3968.

Gomez-Navarro, C., Meyer, J. C., Sundaram, R. S., Chuvilin, A., Kurasch, S., Burghard, M., et al. (2010). Atomic structure of reduced graphene oxide. *Nano Letters, 10,* 1144–1148.

Guo, H., Peng, M., Zhu, Z., & Sun, L. (2013). Preparation of reduced graphene oxide by infrared irradiation induced photothermal reduction. *Nanoscale, 5,* 9040–9048.

Guo, H. L., Wang, X. F., Qian, Q. Y., Wang, F. B., & Xia, X. H. (2009). A green approach to the synthesis of graphene nanosheets. *ACS Nano, 3,* 2653–2659.

Hamilton, C. E., Lomeda, J. R., Sun, Z., Tour, J. M., & Barron, A. R. (2009). High-yield organic dispersions of unfunctionalized graphene. *Nano Letters, 9,* 3460–3462.

He, H., Klinowski, J., Forster, M., & Lerf, A. (1998). A new structural model for graphite oxide. *Chemical Physics Letters, 287,* 53–56.

Huang, L., Liu, Y., Ji, L.-C., Xie, Y.-Q., Wang, T., & Shi, W.-Z. (2011). Pulsed laser assisted reduction of graphene oxide. *Carbon, 49,* 2431–2436.

Hummers, W. S., Jr., & Offeman, R. E. (1958). Preparation of graphitic oxide. *Journal of the American Chemical Society, 80,* 1339.

Jin, S. H., Kim, D. H., Jun, G. H., Hong, S. H., & Jeon, S. (2013). Tuning the photoluminescence of graphene quantum dots through the charge transfer effect of functional groups. *ACS Nano, 7,* 1239–1245.

Ju, J., & Chen, W. (2014). Synthesis of highly fluorescent nitrogen-doped graphene quantum dots for sensitive, label-free detection of Fe (III) in aqueous media. *Biosensors & Bioelectronics, 58,* 219–225.

Kaminska, I., Das, M. R., Coffinier, Y., Niedziolka-Jonsson, J., Sobczak, J., Woisel, P., et al. (2012). Reduction and functionalization of graphene oxide sheets using biomimetic dopamine derivatives in one step. *ACS Applied Materials & Interfaces, 4,* 1016–1020.

Khan, U., O'Neill, A., Lotya, M., De, S., & Coleman, J. N. (2010). High-concentration solvent exfoliation of graphene. *Small, 6,* 864–871.

Khanra, P., Kuila, T., Kim, N. H., Bae, S. H., Yu, D.-S., & Lee, J. H. (2012). Simultaneous bio-functionalization and reduction of graphene oxide by baker's yeast. *Chemical Engineering Journal, 183,* 526–533.

Kim, J., Cote, L. J., & Huang, J. (2012). Two dimensional soft material: New faces of graphene oxide. *Accounts of Chemical Research, 45,* 1356–1364.

Kim, J. H., Chang, W. S., Kim, D., Yang, J. R., Han, J. T., Lee, G. W., et al. (2015). 3D printing of reduced graphene oxide nanowires. *Advanced Materials, 27,* 157–161.

Kim, J., & Suh, J. S. (2014). Size-controllable and low-cost fabrication of graphene quantum dots using thermal plasma jet. *ACS Nano, 8,* 4190–4196.

Krishnamoorthy, K., Veerapandian, M., Yun, K., & Kim, S.-J. (2013). The chemical and structural analysis of graphene oxide with different degrees of oxidation. *Carbon, 53,* 38–49.

Kuila, T., Bose, S., Khanra, P., Mishra, A. K., Kim, N. H., & Lee, J. H. (2012). A green approach for the reduction of graphene oxide by wild carrot root. *Carbon, 50,* 914–921.

Kumawat, M. K., Thakur, M., Bahadur, R., Kaku, T., Prabhuraj, R. S., Ninawe, A., et al. (2019). Preparation of graphene oxide-graphene quantum dots hybrid and its application in cancer theranostics. *Materials Science and Engineering: C, 103,* 109774.

Kumawat, M. K., Thakur, M., Gurung, R. B., & Srivastava, R. (2017a). Graphene quantum dots for cell proliferation, nucleus imaging, and photoluminescent sensing applications. *Scientific Reports, 7,* 15858.

Kumawat, M. K., Thakur, M., Gurung, R. B., & Srivastava, R. (2017b). Graphene quantum dots from mangifera indica: Application in near-infrared bioimaging and intracellular nanothermometry. *ACS Sustainable Chemistry & Engineering, 5,* 1382–1391.

Lei, Z., Lu, L., & Zhao, X. (2012). The electrocapacitive properties of graphene oxide reduced by urea. *Energy & Environmental Science, 5,* 6391–6399.

Lerf, A., He, H., Forster, M., & Klinowski, J. (1998). Structure of graphite oxide revisited. *The Journal of Physical Chemistry B, 102,* 4477–4482.

Li, H., He, X., Kang, Z., Huang, H., Liu, Y., Liu, J., et al. (2010). Water-soluble fluorescent carbon quantum dots and photocatalyst design. *Angewandte Chemie International Edition in English, 49,* 4430–4434.

Li, Y., Hu, Y., Zhao, Y., Shi, G., Deng, L., Hou, Y., & Qu, L. (2011). An electrochemical avenue to green-luminescent graphene quantum dots as potential electron-acceptors for photovoltaics. *Advanced materials, 23*(6), 776–780.

Li, L. L., Ji, J., Fei, R., Wang, C. Z., Lu, Q., Zhang, J. R., et al. (2012). A facile microwave avenue to electrochemiluminescent two-color graphene quantum dots. *Advanced Functional Materials, 22,* 2971–2979.

Li, X., Wang, H., Robinson, J. T., Sanchez, H., Diankov, G., & Dai, H. (2009). Simultaneous nitrogen doping and reduction of graphene oxide. *Journal of the American Chemical Society, 131,* 15939–15944.

Liu, H., Ye, T., & Mao, C. (2007). Fluorescent carbon nanoparticles derived from candle soot. *Angewandte Chemie International Edition in English, 46,* 6473–6475.

Liu, J., Fu, S., Yuan, B., Li, Y., & Deng, Z. (2010). Toward a universal "adhesive nanosheet" for the assembly of multiple nanoparticles based on a protein-induced reduction/decoration of graphene oxide. *Journal of the American Chemical Society, 132,* 7279–7281.

Liu, R., Wu, D., Feng, X., & Mullen, K. (2011). Bottom-up fabrication of photoluminescent graphene quantum dots with uniform morphology. *Journal of the American Chemical Society, 133,* 15221–15223.

Liu, Y., Huang, L., Guo, G. L., Ji, L. C., Wang, T., Xie, Y. Q., et al. (2012). Pulsed laser assisted reduction of graphene oxide as a flexible transparent conducting material. *Journal of Nanoscience and Nanotechnology, 12,* 6480–6483.

Lu, J., Yeo, P. S. E., Gan, C. K., Wu, P., & Loh, K. P. (2011). Transforming C 60 molecules into graphene quantum dots. *Nature Nanotechnology, 6,* 247–252.

Marcano, D. C., Kosynkin, D. V., Berlin, J. M., Sinitskii, A., Sun, Z., Slesarev, A., et al. (2010). Improved synthesis of graphene oxide. *ACS Nano, 4,* 4806–4814.

Moon, I. K., Lee, J., Ruoff, R. S., & Lee, H. (2010). Reduced graphene oxide by chemical graphitization. *Nature Communications, 1,* 73.

Mueller, M. L., Yan, X., McGuire, J. A., & Li, L. S. (2010). Triplet States and electronic relaxation in photoexcited graphene quantum dots. *Nano Letters, 10,* 2679–2682.

Pan, D., Zhang, J., Li, Z., & Wu, M. (2010). Hydrothermal route for cutting graphene sheets into blue-luminescent graphene quantum dots. *Advanced Materials, 22,* 734–738.

Paredes, J. I., Villar-Rodil, S., Martinez-Alonso, A., & Tascon, J. M. (2008). Graphene oxide dispersions in organic solvents. *Langmuir, 24,* 10560–10564.

Pham, V. H., Pham, H. D., Dang, T. T., Hur, S. H., Kim, E. J., Kong, B. S., et al. (2012). Chemical reduction of an aqueous suspension of graphene oxide by nascent hydrogen. *Journal of Materials Chemistry, 22,* 10530–10536.

Radovic, L. R., & Bockrath, B. (2005). On the chemical nature of graphene edges: Origin of stability and potential for magnetism in carbon materials. *Journal of the American Chemical Society, 127,* 5917–5927.

Russier, J., Treossi, E., Scarsi, A., Perrozzi, F., Dumortier, H., Ottaviano, L., et al. (2013). Evidencing the mask effect of graphene oxide: A comparative study on primary human and murine phagocytic cells. *Nanoscale, 5,* 11234–11247.

Salas, E. C., Sun, Z., Lüttge, A., & Tour, J. M. (2010). Reduction of graphene oxide via bacterial respiration. *ACS Nano, 4,* 4852–4856.

Sanchez, V. C., Jachak, A., Hurt, R. H., & Kane, A. B. (2012). Biological interactions of graphene-family nanomaterials: An interdisciplinary review. *Chemical Research in Toxicology, 25,* 15–34.

Seo, M., Yoon, D., Hwang, K. S., Kang, J. W., & Kim, J. (2013). Supercritical alcohols as solvents and reducing agents for the synthesis of reduced graphene oxide. *Carbon, 64,* 207–218.

Shen, J., Zhu, Y., Chen, C., Yang, X., & Li, C. (2011). Facile preparation and upconversion luminescence of graphene quantum dots. *Chemical Communications (Cambridge, England), 47,* 2580–2582.

Shen, J., Zhu, Y., Yang, X., & Li, C. (2012a). Graphene quantum dots: Emergent nanolights for bioimaging, sensors, catalysis and photovoltaic devices. *Chemical Communications (Cambridge, England), 48,* 3686–3699.

Shen, J., Zhu, Y., Yang, X., Zong, J., Zhang, J., & Li, C. (2012b). One-pot hydrothermal synthesis of graphene quantum dots surface-passivated by polyethylene glycol and their photoelectric conversion under near-infrared light. *New Journal of Chemistry, 36,* 97–101.

Shinde, D. B., & Pillai, V. K. (2012). Electrochemical preparation of luminescent graphene quantum dots from multiwalled carbon nanotubes. *Chemistry, 18,* 12522–12528.

Shin, H. J., Kim, K. K., Benayad, A., Yoon, S. M., Park, H. K., Jung, I. S., et al. (2009). Efficient reduction of graphite oxide by sodium borohydride and its effect on electrical conductance. *Advanced Functional Materials, 19,* 1987–1992.

Stankovich, S., Dikin, D. A., Piner, R. D., Kohlhaas, K. A., Kleinhammes, A., Jia, Y., Wu, Y., Nguyen, S. T., & Ruoff, R. S. (2007). Synthesis of graphene-based nanosheets via chemical reduction of exfoliated graphite oxide. *Carbon, 45,* 1558–1565.

Staudenmaier, L. (1898). Verfahren zur darstellung der graphitsäure. *Berichte der Deutschen Chemischen Gesellschaft, 31,* 1481–1487.

Tabrizi, M. A., Azar, S. J., & Varkani, J. N. (2014). Eco-synthesis of graphene and its use in dihydronicotinamide adenine dinucleotide sensing. *Analytical Biochemistry, 460,* 29–35.

Tang, Y., Guo, Q., Chen, Z., Zhang, X., & Lu, C. (2019). In-situ reduction of graphene oxide-wrapped porous polyurethane scaffolds: Synergistic enhancement of mechanical properties and piezoresistivity. *Composites Part A Applied Science and Manufacturing, 116,* 106–113.

Terrones, M., Botello-Méndez, A. R., Campos-Delgado, J., López-Urías, F., Vega-Cantú, Y. I., Rodríguez-Macías, F. J., et al. (2010). Graphene and graphite nanoribbons: Morphology, properties, synthesis, defects and applications. *Nano today, 5,* 351–372.

Tetsuka, H., Asahi, R., Nagoya, A., Okamoto, K., Tajima, I., Ohta, R., et al. (2012). Optically tunable amino-functionalized graphene quantum dots. *Advanced Materials, 24,* 5333–5338.

Tran, D. N., Kabiri, S., & Losic, D. (2014). A green approach for the reduction of graphene oxide nanosheets using non-aromatic amino acids. *Carbon, 76,* 193–202.

Voiry, D., Yang, J., Kupferberg, J., Fullon, R., Lee, C., Jeong, H. Y., et al. (2016). High-quality graphene via microwave reduction of solution-exfoliated graphene oxide. *Science, 353,* 1413–1416.

Wang, L., Wang, Y., Xu, T., Liao, H., Yao, C., Liu, Y., et al. (2014). Gram-scale synthesis of single-crystalline graphene quantum dots with superior optical properties. *Nature Communications, 5,* 5357.

Wang, Y., Shi, Z., & Yin, J. (2011). Facile synthesis of soluble graphene via a green reduction of graphene oxide in tea solution and its biocomposites. *ACS Applied Materials & Interfaces, 3,* 1127–1133.

Wick, P., Louw-Gaume, A. E., Kucki, M., Krug, H. F., Kostarelos, K., Fadeel, B., et al. (2014). Classification framework for graphene-based materials. *Angewandte Chemie International Edition in English, 53,* 7714–7718.

Wijaya, R., Andersan, G., Santoso, S. P., & Irawaty, W. (2020). Green reduction of graphene oxide using kaffir lime peel extract (Citrus hystrix) and its application as adsorbent for methylene blue. *Scientific Reports, 10,* 1–9.

Xue, Q., Huang, H., Wang, L., Chen, Z., Wu, M., Li, Z., et al. (2013). Nearly monodisperse graphene quantum dots fabricated by amine-assisted cutting and ultrafiltration. *Nanoscale, 5,* 12098–12103.

Yin, R., Shen, P., & Lu, Z. (2019). A green approach for the reduction of graphene oxide by the ultraviolet/sulfite process. *Journal of Colloid and Interface Science, 550,* 110–116.

Zhang, M., Bai, L., Shang, W., Xie, W., Ma, H., Fu, Y., et al. (2012). Facile synthesis of water-soluble, highly fluorescent graphene quantum dots as a robust biological label for stem cells. *Journal of Materials Chemistry, 22,* 7461–7467.

Zheng, X. T., Ananthanarayanan, A., Luo, K. Q., & Chen, P. (2015). Glowing graphene quantum dots and carbon dots: Properties, syntheses, and biological applications. *Small, 11,* 1620–1636.

Zhou, X., Zhang, J., Wu, H., Yang, H., Zhang, J., & Guo, S. (2011). Reducing graphene oxide via hydroxylamine: A simple and efficient route to graphene. *The Journal of Physical Chemistry C, 115,* 11957–11961.

Zhu, S., Wang, L., Li, B., Song, Y., Zhao, X., Zhang, G., et al. (2014). Investigation of photoluminescence mechanism of graphene quantum dots and evaluation of their assembly into polymer dots. *Carbon, 77,* 462–472.

Zhu, S., Zhang, J., Tang, S., Qiao, C., Wang, L., Wang, H., et al. (2012). Surface chemistry routes to modulate the photoluminescence of graphene quantum dots: From fluorescence mechanism to up-conversion bioimaging applications. *Advanced Functional Materials, 22,* 4732–4740.

Chapter 3
Physicochemical Properties and Toxicity Analysis

3.1 Introduction

Graphene is a hexagonal crystal of sp^2 hybridized carbon in the form of a two-dimensional material which has shown a multitude of applications. However, the material is limited with respect to its application in devices due to its gapless nature, storage, and processing. These limitation needs to be overcome to bring graphene to industrial level, and scale-up for effective utilization. The primary difference between graphene and graphene oxide (GO) is that the latter contains oxygen which makes it hydrophilic and dispersible in water. This means that graphene oxide contains both sp^2 and sp^3 domains making it susceptible to more interactions. The functional groups present on graphene make it easier to functionalize these materials and their properties to be altered to suit and improve on a particular application. The extraordinary electrical, mechanical, optical properties can be efficiently utilized and enhanced though these functionalization as we will notice in this chapter. The functionalization not only alters the band gap, it also helps enhance the water dispersibility which is of utmost importance in cancer applications. Several procedures have been developed to make dispersible graphene such as chemical vapor deposition, chemical or thermal reduction of GO, intercalative expansion of graphite.

3.2 Covalent Interactions

Covalent modification creates a sp^3 distortion forming a tetrahedral structure due to elongated bond length. Strategies previously employed for functionalization of fullerenes or carbon nanotubes were used to create a similar covalent functionalization with graphene (Fig. 3.1). The covalent functionalization on the graphene structure is carried out based on two essential strategies: (i) Covalent bonds between

R. Srivastava et al., *Next Generation Graphene Nanomaterials for Cancer Theranostic Applications*,
https://doi.org/10.1007/978-981-33-6303-8_3

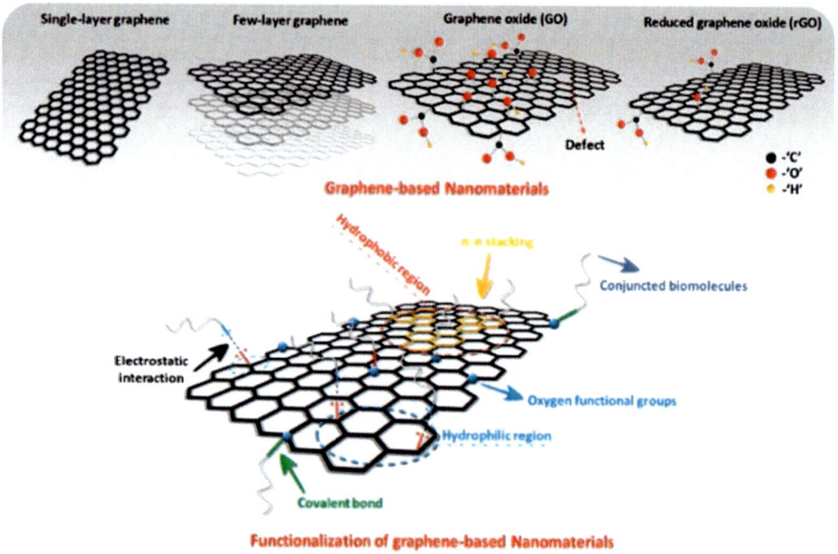

Fig. 3.1 Schematic showing various covalent and non-covalent functionalization of graphene-based nanomaterials. Adapted from Bei et al. (2019)

free radicals and C=C bonds of pristine graphene, (ii) Covalent bonds between organic functional groups and the oxygen functional groups present on the surface of GO (Georgakilas et al. 2012). The covalent modification can occur through four routes: nucleophilic substitution, electrophilic addition, condensation, and addition.

The lone pair of electron on the NH_2 groups attacks the epoxy groups of GO which is the basis for the most promising modification on the GO surface functionality based on nucleophilic substitution reaction. Navaee et al. functionalized GO with –NH2 groups using the well-known Bucherer reaction, through the reaction between ammonia and GO, using sodium bisulfite as the catalyst. Raman spectroscopy can be used to estimate the amount of covalent functionalization as the sp^2 to sp^3 hybridization can be calculated using the I_d/I_g ratio, which gives the defect to graphite content ratio with peaks at 1350 and 1580 cm^{-1}, respectively. The GO forms unstable dispersions in solvents such as water, DMF, THF, NMP as it aggregates through π–π stacking and forms graphite oxide (GO). This can be overcome by using various stabilizing groups known in the literature (Navaee and Salimi 2015). They showed superior electrocatalytic activity and stability toward ORR, and the increased amine functionalities can help in supporting biomolecules, metal nanoparticles which can aide in improving the electrocatalytic properties further. For reinforcement in epoxy resin matrices, a simple temperature-assisted reflux method was used to synthesize GO with n-butylamine, without using any toxic chemicals (Chakraborty et al. 2016) (Fig. 3.2).

In the electrophilic substitution reaction, the hydrogen is replaced by an electrophile. Graphene was iodinated using the electrophilic substitution reaction and

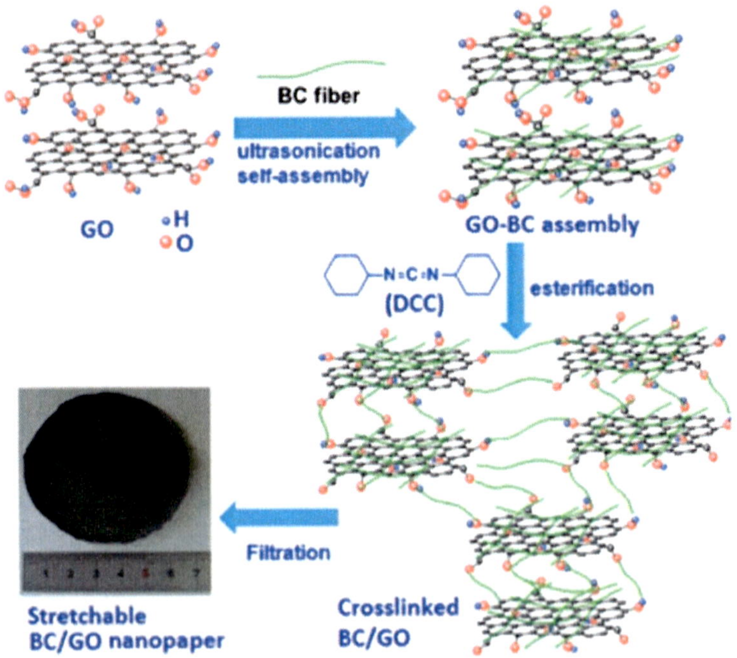

Fig. 3.2 Depicting the covalent bonding between bacterial cellulose and graphene. Adapted from Liu et al. (2015)

graphene oxide was iodinated using nucleophilic substitution. These were observed to favor the electronic transport in fuel cell applications (Marinoiu et al. 2019). Similarly, bromine was covalently bonded to the defect sites (sp^2 C–H) present on the surface of reduced graphene oxide. Covalent functionalization of GO with ferrocene was carried out by Avinash et al. which showed an increase in d-spacing to 2.4 nm, with interesting magnetic properties (Avinash et al. 2010). Using diazonium chemistry, covalent functionalization was carried out on the surface of graphene nanoribbons, which has previously been shown for graphene sheets and single-walled carbon nanotubes. Their good dispersibility in organic solvents can be useful for applications in composites and thin film materials (Zhu et al. 2009). Through effective reductive activation, charged graphene was functionalized using organic diazonium salts. Bruce et al. (Jonathan and Jude 2019) functionalized formate esters on the surface of graphene quantum dots through electrophilic substitution, which showed a red-shift coupled with a reduction in band gap. These can be utilized for single photon emission, tunable fluorescence for optical and cellular imaging among others (Fig. 3.3).

Condensation reaction occurs when two molecules (functional groups) combine to form a single molecule with a loss of entropy. In the case of graphene, condensation occurs through oxygenated groups with isocyanate, diisocyanate, and

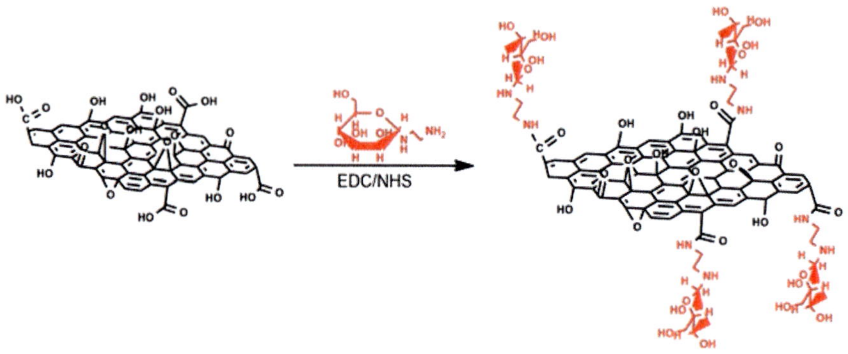

Fig. 3.3 Functionalization of GO with mannosylated ethylenediamine using EDC and NHS crosslinking agents Taken with permission from de Sousa et al. (2018)

amine compounds through an amidation reaction. Ethylenediamine was used a solid base catalyst to functionalize graphene oxide for Knoevenagel condensation reaction (Xue et al. 2015). Similarly, PdNi were furnished on the GO as a heterogeneous catalyst for the promotion of Knoevenagel reaction between malononitrile and aromatic aldehydes (Lolak et al. 2019). Through Claisen-Schmidt condensation reaction hierarchal mesoporous MgO is synthesized in the presence of graphene by a simple hydrothermal homogenous precipitation method. In this procedure, Qi et al. (2019) have used urea as the precipitating reagent. Graphene oxide sheets were covalently pillared with metallomacrocycles, to form a porous framework with tunable porosity with Ni centers (Kim et al. 2013). The first example of condensation through amine terminated poly (ethylene glycol) was reported by Liu et al. which was used for delivery of hydrophobic anticancer drugs (Liu et al. 2008) (Fig. 3.4).

The covalent functionalization through addition reaction of reduced graphene oxide was carried out by conjugated polymers through nucleophilic addition of nitrogen atoms. Three conjugated polymers namely tetraphenylethylene, carbazole, and phenyl groups were used which with the polymer-wrapping effect greatly improved the stability of RGO (Xu et al. 2017). Various polymerization reactions have been used to covalently polymerize the graphene surface such as ring-opening polymerization, Ziegler-Natta polymerization, ATRP method, reversible addition-fragmentation reaction. Polyglycerol was covalently grafted on to the surface of graphene nanosheets through in situ ring-opening polymerization of glycidol. Park and Yan (2013) showed the graphene functionalization focusing on reactive intermediates of radicals, nitrenes, carbenes, and arynes. The reactive species form covalent adducts with sp^2 carbon structures in graphene. The covalent bond formation was established by XPS. Zhang et al. (2013) synthesized Janus like graphene using covalent functionalization further assisted by poly (methyl methacrylate)-mediated transfer approach.

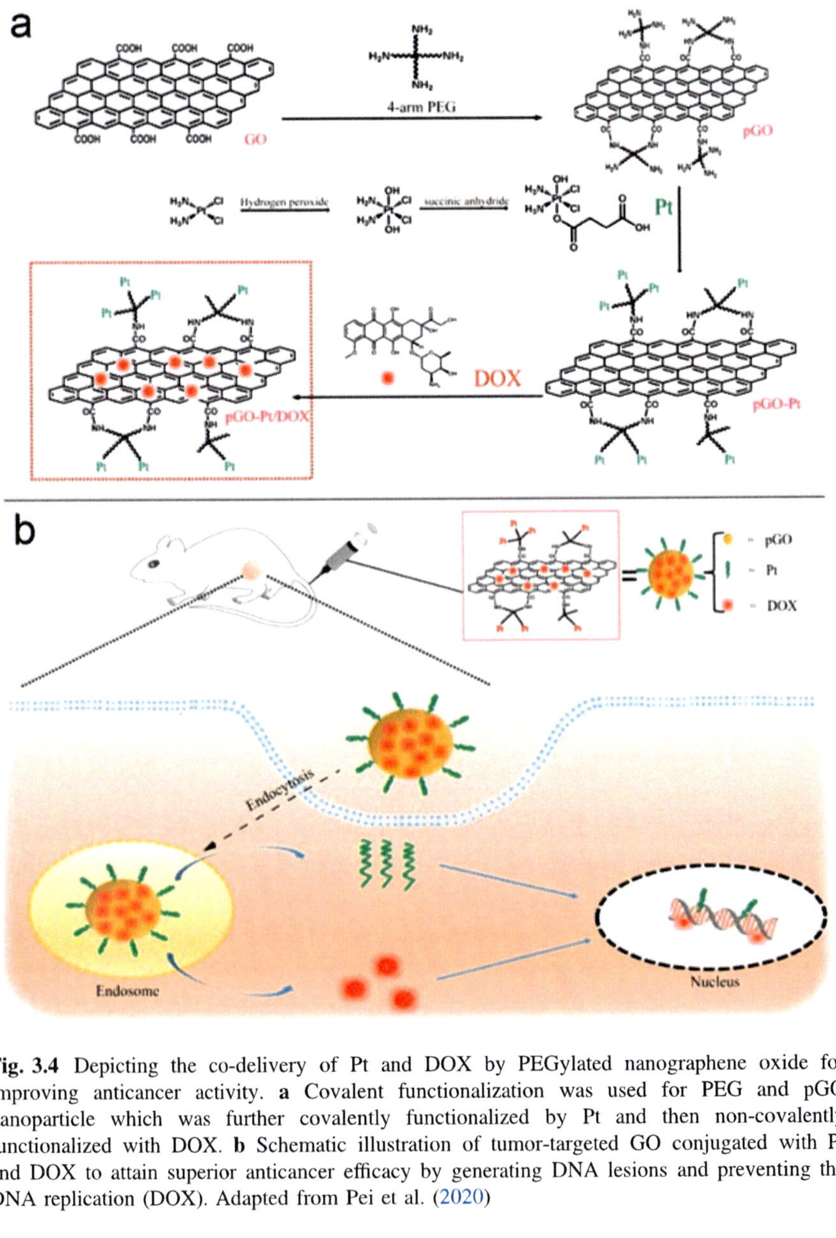

Fig. 3.4 Depicting the co-delivery of Pt and DOX by PEGylated nanographene oxide for improving anticancer activity. **a** Covalent functionalization was used for PEG and pGO nanoparticle which was further covalently functionalized by Pt and then non-covalently functionalized with DOX. **b** Schematic illustration of tumor-targeted GO conjugated with Pt and DOX to attain superior anticancer efficacy by generating DNA lesions and preventing the DNA replication (DOX). Adapted from Pei et al. (2020)

3.3 Non-covalent Interactions

Covalent modifications result in destroying the intrinsic graphene conjugation systems, therefore non-covalent modifications, i.e., Van der Waals forces, π–π stacking interactions are carried out. Van der Waals forces are electrostatic

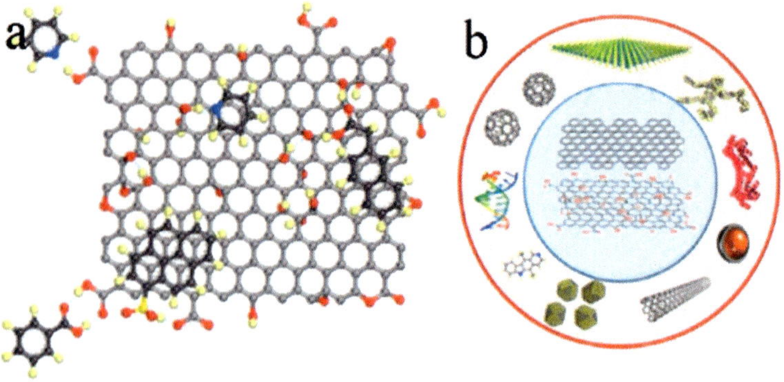

Fig. 3.5 **a** Hydrogen bonding and/or π–π stacking with GO, **b** selected non-covalent interactions of graphene and graphene oxide with selected species such as MoS_2, aromatic polymers, proteins, core-shell nanoparticle, nanotube, quantum dots, ellipticine, (drug/photoluminescence probe), DNA segment, and fullerenes. Adapted from Georgakilas et al. (2016)

interactions involving permanent or induced dipoles. Similarly, the interaction of the π orbital in the orbital system ensues π–π interactions. These modifications help in retaining the material properties while giving them interesting modifications. Non-covalent polymer functionalization of graphene can occur through hydrogen bonding and π–π stacking interactions which include polymers with pyrene-end groups, conjugated polyelectrolytes, melt blending (Fig. 3.5).

For the π–π interactions, the existence of π systems as well as the geometry has to be considered. The aromaticity is involved in the π–π interaction. The non-covalent functionalization on the basal plane of graphene with the π system of alkyne-functionalized pyrene was carried out by Shen et al. (2013). This led to the formation of bifunctional modular unit of alkyne-functionalized pyrene. The cation-π interactions of graphene containing systems showed an enhanced membrane stability, selectivity, and adsorption in water treatment systems (Zhao and Zhu 2020). Zhang et al. (2011) discussed the tailoring of bandgap of graphene by non-covalent stacking with aromatic molecules through π–π interactions. Karthik et al. (2018) showed how the π–π interaction between MOF and RGO gives an increased visible-light photocatalytic activity. Results from density functional theory by Su et al. (2011) simulated the interactions between benzene and graphene. This model can be used for the study between interactions of aromatic molecule and graphene. Glucose oxidase, which is an enzyme specific for glucose sensing, was incorporated on the surface of graphene through π–π interaction (Fig. 3.6).

A UV photodetector synthesized by Zhang et al. (2015) shows that the ZnO nucleates at defect sites in graphene. Kim et al. (2015) fabricated a combinatorial pattern of graphene-hybrid to control the differentiation of human adipose-derived mesenchymal stem cells. Extracellular matrix adsorbed on the surface of graphene oxide was shown to increase the therapeutic efficacy of mesenchymal stem cells (Park et al. 2015). Graphene oxide functionalized on a patterned gold surface was

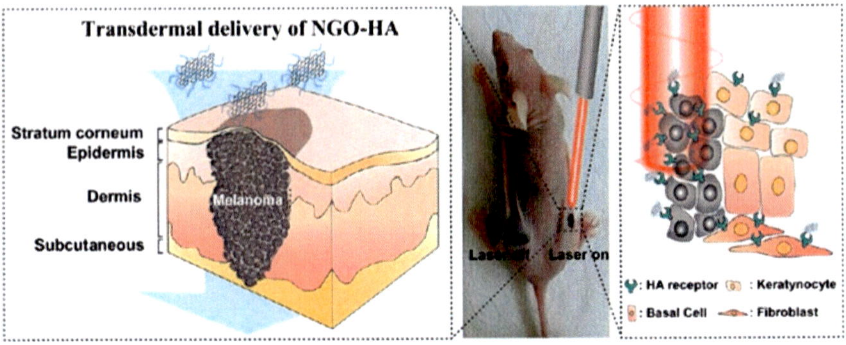

Fig. 3.6 Schematic illustration of nanographene oxide-hyaluronic acid (NGO-HA) conjugate transdermal delivery into melanoma skin cancer cells followed by photothermal therapy using NIR laser. Adapted from Jung et al. (2014)

used in the form of a microfluidic device to isolate circulating tumor cells from lung, breast, and pancreatic cancer patients (Yoon et al. 2013). Polyethylene glycerol which was non-covalently functionalized on reduced graphene oxide exhibited six-fold higher NIR absorption than the RGO that was covalently functionalized with PEG. The NIR absorption property of these PEGylated RGO was used for photothermal therapy of U87MG cancer cells in vitro (Robinson et al. 2011). Zhang et al. (2010) used both covalent and non-covalent interactions for targeting and drug loading respectively. Firstly, folic acid which is used to target MCF-7 breast cancer cells was covalently binded with the sulfonic acid groups functionalized on the surface of graphene oxide. Further, doxorubicin and camptothecin, two anticancer drugs for the combinatorial therapy, were conjugated on the surface of this material via π–π stacking and hydrophobic interactions (Fig. 3.7).

Liquid phase exfoliation of graphene is carried out by exfoliation of graphite, followed by stabilization of the graphene sheets. Finally, the bulk graphitic material is removed through various purifications steps such as filtration or centrifugation. For the successful exfoliation, the van der Waals interaction should be overcome by external physical forces. Exfoliation through ultrasonication is a low-cost, green strategy to obtain highly dispersible graphene which has been shown to be effective for h-BN, MoS_2, WS_2, and other 2D materials. The π–π interactions were studied through single molecule force spectroscopy by Zhang et al. (2007). To sum up, the covalent as well as non-covalent bonding in the graphene structures instilled in these materials interesting properties which help in overcoming the limitations of graphene-based materials such as fixed band gap, water dispersibility, and at the same time enhancing some of its properties for various applications. While covalent interactions are stronger forms of interactions which can create a change in the inherent graphene structure, non-covalent preserve the graphene structure and at the same time give them interesting modifications which show some exciting

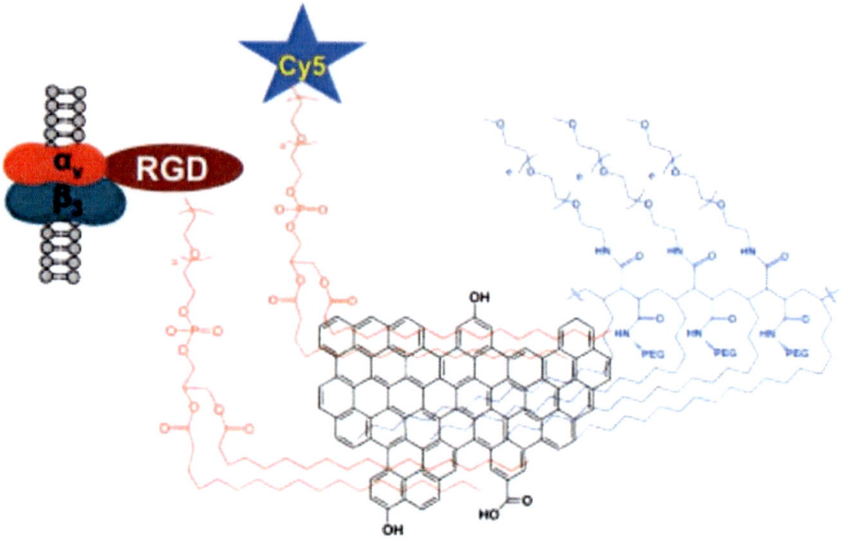

Fig. 3.7 Showing the interaction of RGO with the $\alpha_v\beta_3$ receptors of U87MG cell. Adapted from Robinson et al. (2011)

properties. These kinds of interactions along with being cost-effective make graphene an exceptional 2D material which has resulted in exploiting these for a multitude of applications.

3.4 Modification and Their Implications

The graphene-based nanomaterials can be modified either post-synthesis (extrinsic modification) or during synthesis (intrinsic modification). The major reasons for surface modifications (or functionalization) could be either to tune the intrinsic optical photoluminescence, enabling chemical stability and biocompatibility, and enable further functionalization with chemotherapeutics drugs for drug delivery applications.

3.4.1 Extrinsic Modification

In the search for enhancing the properties of the graphene-based materials, few attempts have been made using extrinsic modification using organic molecules. For instance, Yang et al., used an amino molecule—ethoxypropylamine (EPA) for extrinsic modification of carbon quantum dots (Fig. 3.8a) and demonstrated

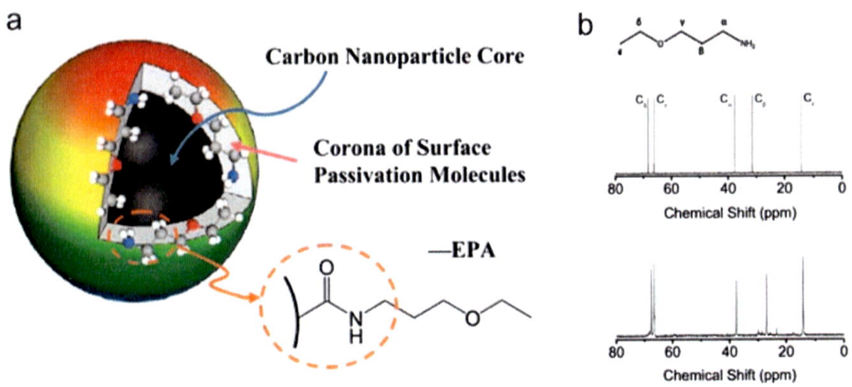

Fig. 3.8 **a** Schematic showing external modification of Carbon quantum dots using 3-ethoxypropylamine (EPA) for enhancement of fluorescence emission. **b** Proton NMR showing successful functionalization of nanodots with EPA. Reprinted (adapted) with permission from Yang et al. (2016). Copyright (2016) American Chemical Society

progress in optical performance. The functionalization is formed by amide linkages between the exposed carboxylic group from the carbon quantum dots and EPA molecule. They observed that post-functionalization with EPA (Fig. 3.8b), there was a new optical absorption peak around 400–450 nm which enhances the green fluorescence emission from the nanodots. The reasons for fluorescence enhancement could be following either significant contributions associated with enhanced transition probabilities or due to the formation of the emissive excited states and their corresponding decays. The implication of this study is that this approach could be used to enhance optical bioimaging of cancer diagnosis.

Another most widely used functionalization molecule is polyethylene glycol (PEG) for making carbon nanoparticles compatible with in vivo biological studies. The functionalization procedure is referred to as PEGylation and is often done to enhance the stability and biocompatibility of the molecule. Sun and co-workers explored PEGylation CDs with PEG_{1500N} and observed that the CDs were non-toxic (in vitro and in vivo) (Yang et al. 2009; Wang et al. 2010).

In another study, other molecules were explored by Qian et al., for functionalizing graphene quantum dots (GQDs) (Qian et al. 2013). They functionalized GQDs with 1,2-glycol, 1,2-dithioglycol, 1,2-ethylenediamine, 1,3-propanediamine, 1,4-butanediamine, and ethylamine. The idea was to enhance the quantum yield of GQDs followed by changing the surface chemistry of the GQDs. The GQDs were synthesized from graphitic powder and then surface functionalized with these representative molecules (amines, diamines, dialcohols, dithiols). They observed that the amino functionalization greatly improved the quantum yield ($\sim$17%) and depends upon the length of the amino chains.

One of the interesting applications of extrinsic modification of CDs would be to impart a novel characteristic to the structure which may be triggered by changes in the extracellular environment. One such study was shown by Feng et al., wherein they prepared a complex structure of CDs and functionalized with Pt(IV)@PEG-(PAH/DMMA) (Feng et al. 2016). The anionic polymer along with dimethylmaleic acid (PEG-(PAH/DMMA) has a charge-convertible characteristic in mildly acidic environment. Since tumor microenvironment is acidic in nature, such molecule will undergo charge conversion and release drug (Pt(IV)) from the complex (Fig. 3.9).

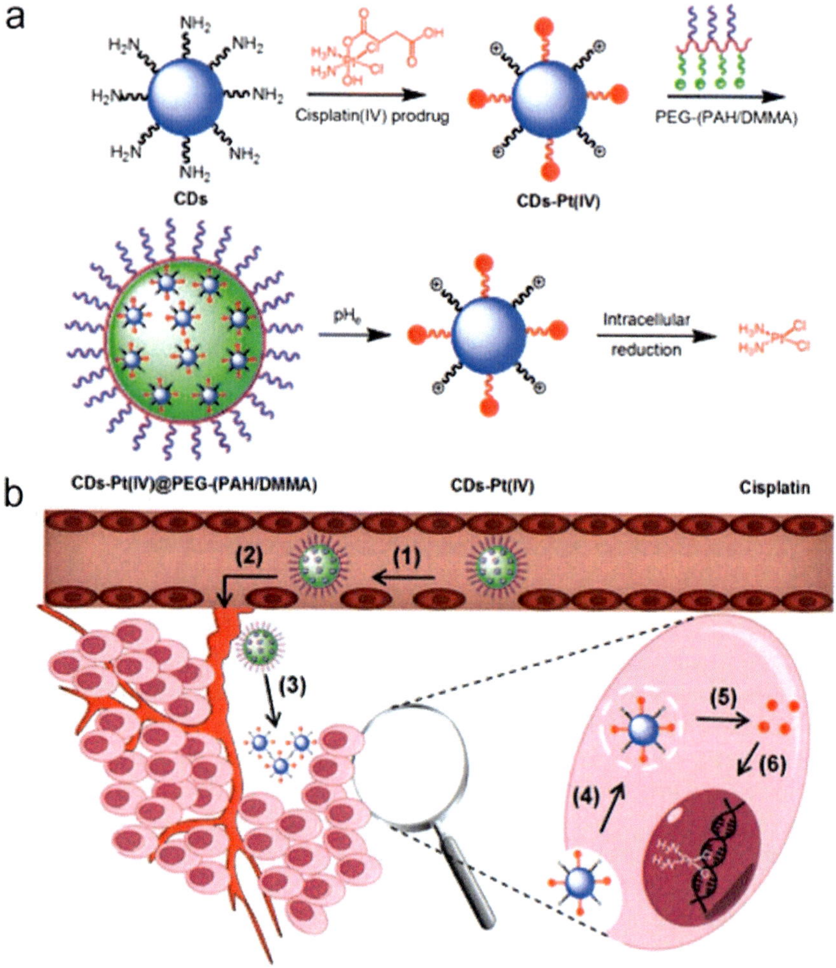

Fig. 3.9 **a** Cartoon showing functionalization of CDs with chemotherapeutic drug (Cisplatin) and PEG-(PAH/DMMA). **b** Effect of functionalization in stimuli-responsive drug delivery wherein the negatively charged CDs (owing to PEGylation) render increased blood residence time, followed by flushing in the tumor site due to the EPR effect, and conversion of charge at tumor pH releasing the drug. The drug is then ingested in the cancer cell where the prodrug is converted into active drug followed by its effect on DNA damage and thus killing the cancer cell. Reprinted (adapted) with permission from Feng et al. (2016). Copyright (2016) American Chemical Society

Other than organic molecules, few strategies have been focused upon the synergistic properties of two or more carbon nanoforms (e.g., GQDs on GO—a hybrid of 2D-0D systems). Graphene-based oxides are inherently poorly fluorescent material. To realize their application in diagnostics, theranostics, and biomedical optics, one needs to conjugate these nanomaterials with an organic dye/imaging probe to locate them in the extra/intracellular environment. The main concern of these dyes is their photo-instability, blinking, and sensitivity to the environment, which reduces their efficiency and undermines the purpose.

To address these issues, Srivastava and co-workers use GQDs, a zero-dimensional derivative of GO which bears excellent physicochemical properties like excellent fluorescence property, photostability and biocompatibility (Kumawat et al. 2019). The GQDs are easy to synthesize using various carbon-based precursors and thereby widely employed in bioimaging and photothermal response. Taking advantage of these properties of GQDs, they functionalized the GO sheets with GQDs using polyethyleneimine (PEI) as a linker (Figs. 3.10 and 3.11). The PEI is a cationic polymer used for surface modification of graphene nanomaterials, linking of two nano-constructs or moieties, chemical scissoring, biosensing, gene delivery and bioimaging, and fluorescence enhancing applications. Since PEI is positively charged, it helps in bringing two negatively charged moieties together and make a stable hybrid. The hybrid was used for cellular imaging, photothermal therapy, and oxidative stress generation for cancer therapy.

3.4.2 *Intrinsic Modification*

Intrinsic modification refers to an *in situ* modification of graphene-based nanomaterials during the synthesis of the material itself. Usually these processes involve addition of one or two elements in the nanomaterial's basal carbon structure or intercalation, thereby enhancing the optical properties, also called as doping. Doping is the most common method to modify carbon nanomaterials, especially optically active ones such as CDs.

Heteroatom doping is widely known method of tuning the optical and physical properties of CDs. The photoluminescence property of CDs, for instance, can be tuned by doping. Doping introduces an allowed state in the band gap of the materials between highest occupied molecular orbital (HUMO) and lowest unoccupied molecular orbital (LUMO). Dopants with higher electronegativity than carbon (e.g., N-doping) cause electron donating effect and blue-shift in the PL of CDs (Fig. 3.12). On the contrary, dopants with lower electronegativity cause red-shift due to electron donating effect, B-doping (Fig. 3.13).

Nitrogen-doping (N-type doping) is one of the ways to tune optical PL properties of CDs and GQDs. Li et al. observed that nitrogen-doping in GQDs via solution chemistry greatly enhanced their optical PL property (Li et al. 2012). The method controlled the dopant concentration in the structure and bonding structure in GQDs.

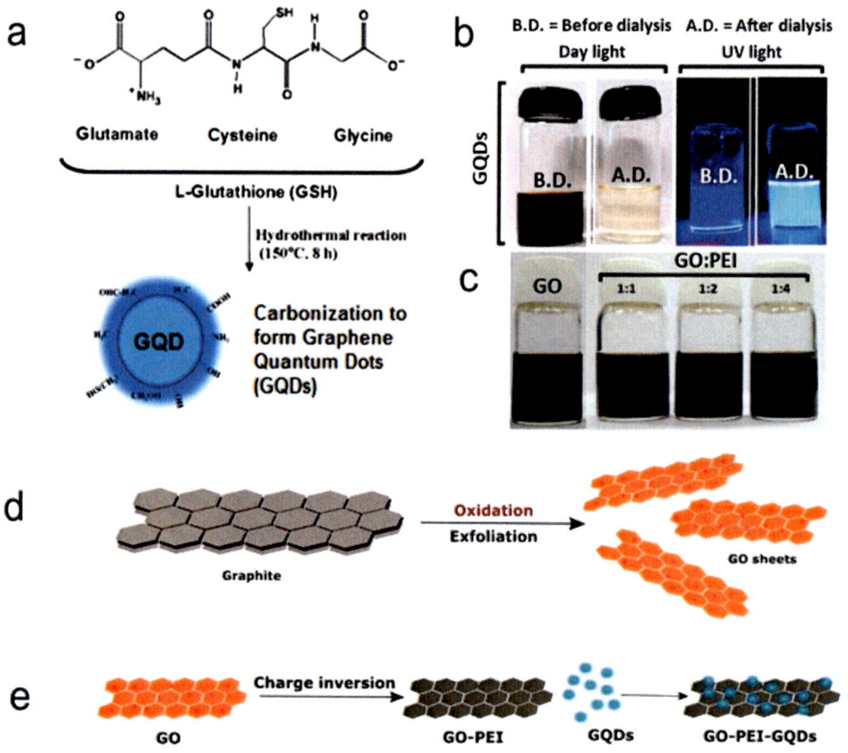

Fig. 3.10 Synthesis of GQDs and formation of GO-PEI-GQDs hybrid. **a** Schematic showing synthesis of GQDs through a hydrothermal carbonization process, **b** Optical images showing GQDs (before and after dialysis) under daylight and UV lamp (375 nm), **c** Stability of dispersed GO and GO: PEI solution at various ratios (w/w). **d** Synthesis of GO and its exfoliation. **e** Functionalization of GO sheets using PEI to form GO-PEI. The GQDs were then functionalized onto GO-PEI sheets to form GO-PEI-GQDs using layer-by-layer electrostatic assembly. Adapted from Kumawat et al. (2019)

Similarly, another method can be used for achieving N-doping in CDs via degradation-based reaction (hydrothermal, microwave heating, etc.). In an attempt to produce such CDs, Liu et al. used a green carbon precursor (Pear juice extract) to produce N-CDs via simple microwave-assisted heating (Liu et al. 2019). They used ethanediamine as an external nitrogen donating source (Fig. 3.14). They observed enhancement in the fluorescence in N-CDs that was then used for biosensing of aptamers in Turn-on/off assay. Another type of doping includes relatively larger atoms in comparison with carbon such as phosphorus and sulfur. Doping with phosphorous and sulfur renders irregularities in the CDs thereby causing disorder in the structure as observed in CNTs (Maciel et al. 2009). Wang et al. prepared water-soluble P-doped CDs (PCDs) using phytic acid (phosphorous rich) carbon source (Wang et al. 2014). The PCDs were highly green fluorescent with quantum yield approaching $\sim 19\%$ and were used for bioimaging of L929 cells.

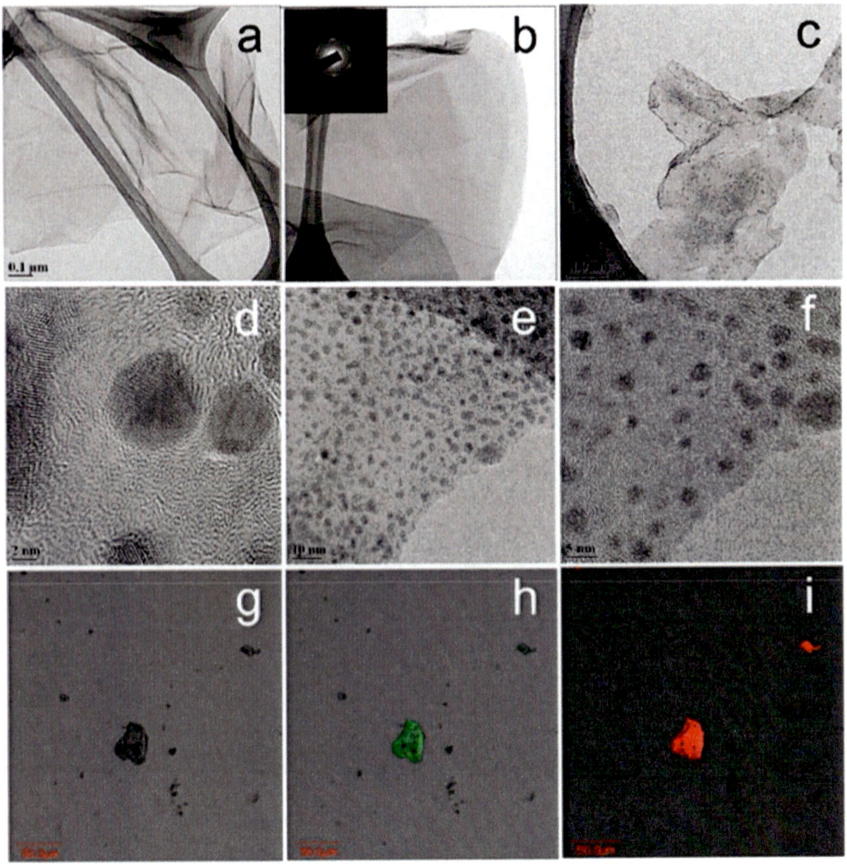

Fig. 3.11 Morphological analysis after functionalization of GO with GQDs using transmission electron microscope. **a** Bare GO sheet, **b** GO sheets functionalized with PEI, **c** GO-PEI-GQDs, **d–f** high-resolution imaging of GO-PEI-GQDs, **g–h** confocal imaging of the functionalized sheets showing multi-color emission due to GQDs. Adapted from Kumawat et al. (2019)

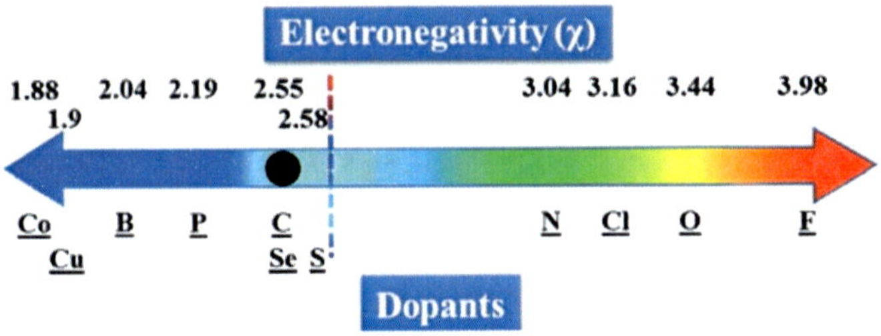

Fig. 3.12 Various dopants used for heteroatom modification of graphene-based nanomaterials. Adapted from Wang et al. (2016)

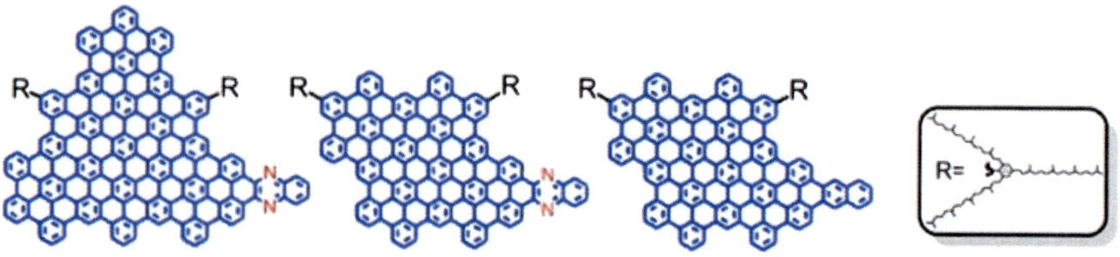

1 2 3

Fig. 3.13 Nitrogen-doping of GQDs (referred to as N-GQDs) 1 and 2 while 3 is undoped GQD structure. Reprinted (adapted) with permission from Li et al. (2012). Copyright (2012) American Chemical Society

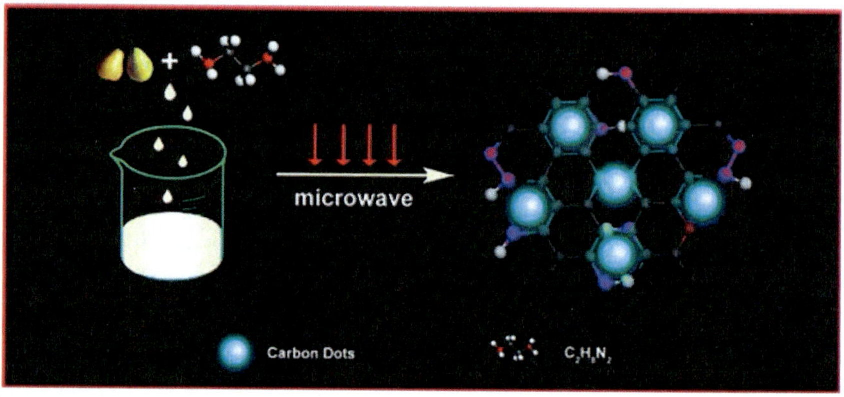

Fig. 3.14 Nitrogen-doping of CDs (referred to as N-CDs) using microwave-assisted heating (400 W, 10 min) using pear juice as carbon and ethanediamine as nitrogen source. Adapted from Liu et al. (2019)

P-type doping constitutes impurities which can act as electron acceptors while carbon basal structure as donors. Boron doping is a typical example of P-type doping and several groups have worked toward modifying GQDs with B-doping. Figure 3.15 shows synthesis of B-doped GQDs using graphite as a carbon source and borax as a boron source (Hai et al. 2015). The emerging B-doped GQDs showed excitation-dependent fluorescence and high biocompatibility (>85% cell viability) even at higher dose (4 mg/mL). Thus, B-doping enhanced the optical activity of GQDs (QY: 21%) and photostability with better biocompatibility.

Co-doping with two or more heteroatoms is another exciting approach for exploring optical properties of graphene-based materials. For instance, co-doping CDs with N and P can be used for biosensing of heavy metal ions (Fe^{3+}) with higher affinity (Shangguan et al. 2017). For this, they used adenosine 5′-triphosphate (ATP) as a single carbon, nitrogen, and phosphorus source. Following hydrothermal degradation, they obtained CDs with QY of ~43%, as well as excellent photostability, low toxicity, and aqueous solubility. Due to co-doping

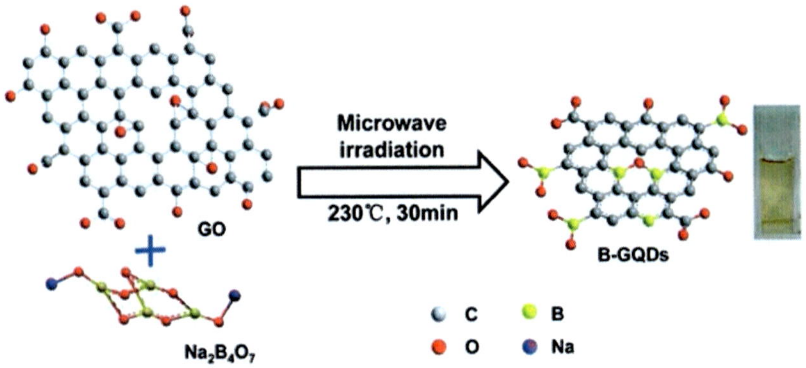

Fig. 3.15 Schematic showing B-doping of GQDs using one-step microwave-assisted heating. Adapted from Hai et al. (2015)

(N/P codoped CDs), Fe–O–P bonds formed between Fe^{3+} and CDs, a very high selectivity was achieved toward Fe^{3+} ions. Co-doping therefore can be used to modulate specific chemical sensing of ions.

3.5 Toxicity Issues

The toxicity of graphene-based materials can be attributed to the chemical response of the nanostructures in the surrounding environment. With different methods of preparation of such carbon nanomaterials, the toxicity differs significantly and depends upon the choice of precursors and additive chemicals used. Extrinsic and intrinsic modifications are therefore one of the ways to tackle such toxicity related issues. Nevertheless, it is important to discuss any toxic effect of such graphene-based nanomaterials that are routinely focused upon for bioimaging and therapy of cancer.

For instance, GQDs are generally shown to be inherently non-toxic and biocompatible for cells. However, there are different aspects to this and should be considered well before generalizing as biocompatible nanomaterial. The cellular toxicity of GQDs (Fig. 3.16) has been attributed to its synthesis method, size, surface functionalization, doping and therapeutic dose, as extensively reviewed by (Wang et al. 2016). Synthesis methods (top-down and bottom-up) for GQDs are important to impart toxicity in bio-applications. For example, top-down approaches involve oxidative degradation of GO, electrochemical reaction, laser-based heating while bottom-up method involves hydrothermal reaction, microwave-assisted heating, or pyrolysis. The former methods usually involve use of toxic chemicals such as concentrated acids (nitric acid), oxidative or corrosive chemicals which might be difficult to purify causing detrimental effects in cells.

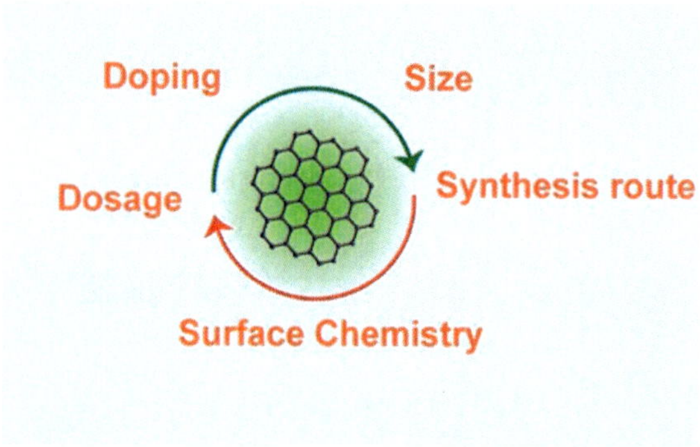

Fig. 3.16 Different parameters that contribute to the toxicity of graphene-based nanomaterials such as GQDs

Size is another factor that might contribute to the cytotoxicity. For instance, larger-sized nanoparticles are difficult for the cells to uptake and produce higher reactive oxygen species (ROS). For example, in terms of size, GO is few micron-sized 2D materials, while GQDs are 1–2 orders of magnitude smaller than GO. In this respect, it is easier for cellular uptake of GQDs rather than GO (Wang et al. 2016). The GO and other forms such as carbon nanotubes (CNTs) have been shown to have detrimental effect on the growth of plant cells (Li et al. 2014). However, few other researchers have argued that even GQDs (>10 nm) are capable of producing ROS species (Markovic et al. 2012). Srivastava and co-workers also observed this activity in GQDs prepared from different sources and showed that such GQDs can be used for photodynamic therapy of cancer (Thakur et al. 2017; Kumawat et al. 2019). It has also been observed that this toxicity can be dose-dependent (Thakur et al. 2017; Kumawat et al. 2019). Typically, a dose of $<200\ \mu g\ mL^{-1}$ is sufficient for a cell viability of $\sim 80\%$ or more (Wang et al. 2016).

The structural chemistry also determines their toxicity in cells. Various functionalization as discussed in the above sections can lead to varied cellular toxicity of GQDs. For example, GQDs functionalized with different functional groups—NH_2, –COOH, and –CO–N $(CH_3)_2$ showed differential cell viability on human neural glioma C6 cells ($200\ \mu g\ mL^{-1}$) (Yuan et al. 2014). Furthermore, chirality of molecule is also shown to affect toxicity. Suzuki et al. functionalized GQDs with two chiral molecules: L/D-cysteine molecules (Suzuki et al. 2016). They observed that D-GQDs interact better to the cell membranes in comparison with L-GQDs and therefore the latter performs relatively better in terms of biocompatibility. Doping of GQDs with heteroatoms such as nitrogen or boron has been shown to improve biocompatibility of GQDs at concentrations >1 mg/mL (Wang et al. 2016).

In general, the toxicity studies from GQDs, CDs are mostly based on in vitro and in vivo studies. Better correlative studies are needed as shown in Fig. 3.16 that imparts biological toxicity of these graphene-based nanomaterials.

References

Avinash, M. B., Subrahmanyam, K. S., Sundarayya, Y., & Govindaraju, T. (2010). Covalent modification and exfoliation of graphene oxide using ferrocene. *Nanoscale, 2,* 1762–1766.

Bei, H. P., Yang, Y., Zhang, Q., Tian, Y., Luo, X., Yang, M., et al. (2019). Graphene-based nanocomposites for neural tissue engineering. *Molecules, 24,* 658.

Chakraborty, S., Saha, S., Dhanak, V. R., Biswas, K., Barbezat, M., Terrasi, G. P., et al. (2016). High yield synthesis of amine functionalized graphene oxide and its surface properties. *RSC Advances, 6,* 67916–67924.

de Sousa, M., Martins, C. H. Z., Franqui, L. S., Fonseca, L. C., Delite, F. S., Lanzoni, E. M., et al. (2018). Covalent functionalization of graphene oxide with d-mannose: Evaluating the hemolytic effect and protein corona formation. *Journal of Materials Chemistry B, 6,* 2803–2812.

Feng, T., Ai, X., An, G., Yang, P., & Zhao, Y. (2016). Charge-convertible carbon dots for imaging-guided drug delivery with enhanced in vivo cancer therapeutic efficiency. *ACS Nano, 10,* 4410–4420.

Georgakilas, V., Otyepka, M., Bourlinos, A. B., Chandra, V., Kim, N., Kemp, K. C., et al. (2012). Functionalization of graphene: Covalent and non-covalent approaches, derivatives and applications. *Chemical Reviews, 112,* 6156–6214.

Georgakilas, V., Tiwari, J. N., Kemp, K. C., Perman, J. A., Bourlinos, A. B., Kim, K. S., et al. (2016). Noncovalent functionalization of graphene and graphene oxide for energy materials, biosensing, catalytic, and biomedical applications. *Chemical Reviews, 116,* 5464–5519.

Hai, X., Mao, Q.-X., Wang, W.-J., Wang, X.-F., Chen, X.-W., & Wang, J.-H. (2015). An acid-free microwave approach to prepare highly luminescent boron-doped graphene quantum dots for cell imaging. *Journal of Materials Chemistry B, 3,* 9109–9114.

Jonathan, B., & Jude, C. (2019). *Facile Synthesis of Formate-Functionalized Graphene Quantum Dots.*

Jung, H. S., Kong, W. H., Sung, D. K., Lee, M.-Y., Beack, S. E., Keum, D. H., et al. (2014). Nanographene oxide-hyaluronic acid conjugate for photothermal ablation therapy of skin cancer. *ACS Nano, 8,* 260–268.

Karthik, P., Vinoth, R., Zhang, P., Choi, W., Balaraman, E., & Neppolian, B. (2018). π–π interaction between metal-organic framework and reduced graphene oxide for visible-light photocatalytic H2 production. *ACS Applied Energy Materials, 1,* 1913–1923.

Kim, T.-H., Shah, S., Yang, L., Yin, P. T., Hossain, M. K., Conley, B., et al. (2015). Controlling differentiation of adipose-derived stem cells using combinatorial graphene hybrid-pattern arrays. *ACS Nano, 9,* 3780–3790.

Kim, T. K., Cheon, J. Y., Yoo, K., Kim, J. W., Hyun, S.-M., Shin, H. S., et al. (2013). Three-dimensional pillared metallomacrocycle–graphene frameworks with tunable micro- and mesoporosity. *Journal of Materials Chemistry A, 1,* 8432–8437.

Kumawat, M. K., Thakur, M., Bahadur, R., Kaku, T., Prabhuraj, R. S., Ninawe, A., et al. (2019). Preparation of graphene oxide-graphene quantum dots hybrid and its application in cancer theranostics. *Materials Science and Engineering C, 103,* 109774.

Li, Q., Zhang, S., Dai, L., & Li, L.-S. (2012). Nitrogen-doped colloidal graphene quantum dots and their size-dependent electrocatalytic activity for the oxygen reduction reaction. *Journal of the American Chemical Society, 134,* 18932–18935.

Li, X., Zhou, Z., Lu, D., Dong, X., Xu, M., Wei, L., et al. (2014). The effect of pristine carbon-based nanomaterial on the growth of green gram sprouts and pH of water. *Nanoscale Research Letters, 9,* 583.

Liu, Y., Jiang, L., Li, B., Fan, X., Wang, W., Liu, P., et al. (2019). Nitrogen doped carbon dots: Mechanism investigation and their application for label free CA125 analysis. *Journal of Materials Chemistry B, 7,* 3053–3058.

Liu, Y., Zhou, J., Zhu, E., Tang, J., Liu, X., & Tang, W. (2015). Facile synthesis of bacterial cellulose fibres covalently intercalated with graphene oxide by one-step cross-linking for robust supercapacitors. *Journal of Materials Chemistry C, 3,* 1011–1017.

Liu, Z., Robinson, J. T., Sun, X., & Dai, H. (2008). PEGylated nanographene oxide for delivery of water-insoluble cancer drugs. *Journal of the American Chemical Society, 130,* 10876–10877.

Lolak, N., Kuyuldar, E., Burhan, H., Goksu, H., Akocak, S., & Sen, F. (2019). Composites of palladium-nickel alloy nanoparticles and graphene oxide for the knoevenagel condensation of aldehydes with malononitrile. *ACS Omega, 4,* 6848–6853.

Maciel, I. O., Campos-Delgado, J., Cruz-Silva, E., Pimenta, M. A., Sumpter, B. G., Meunier, V., et al. (2009). Synthesis, electronic structure, and raman scattering of phosphorus-doped single-wall carbon nanotubes. *Nano Letters, 9,* 2267–2272.

Marinoiu, A., Raceanu, M., Carcadea, E., Varlam, M., & Stefanescu, I. (2019). Iodinated carbon materials for oxygen reduction reaction in proton exchange membrane fuel cell. Scalable synthesis and electrochemical performances. *Arabian Journal of Chemistry, 12,* 868–880.

Markovic, Z. M., Ristic, B. Z., Arsikin, K. M., Klisic, D. G., Harhaji-Trajkovic, L. M., Todorovic-Markovic, B. M., et al. (2012). Graphene quantum dots as autophagy-inducing photodynamic agents. *Biomaterials, 33,* 7084–7092.

Navaee, A., & Salimi, A. (2015). Efficient amine functionalization of graphene oxide through the Bucherer reaction: An extraordinary metal-free electrocatalyst for the oxygen reduction reaction. *RSC Advances, 5,* 59874–59880.

Park, J., Kim, B., Han, J., Oh, J., Park, S., Ryu, S., et al. (2015). Graphene oxide flakes as a cellular adhesive: Prevention of reactive oxygen species mediated death of implanted cells for cardiac repair. *ACS Nano, 9,* 4987–4999.

Park, J., & Yan, M. (2013). Covalent functionalization of graphene with reactive intermediates. *Accounts of Chemical Research, 46,* 181–189.

Pei, X., Zhu, Z., Gan, Z., Chen, J., Zhang, X., Cheng, X., et al. (2020). PEGylated nano-graphene oxide as a nanocarrier for delivering mixed anticancer drugs to improve anticancer activity. *Scientific Reports, 10,* 2717.

Qi, X., Yan, X., Peng, W., Zhang, J., Tong, Y., Li, J., et al. (2019). Graphene-induced hierarchical mesoporous MgO for the Claisen-Schmidt condensation reaction. *New Journal of Chemistry, 43,* 4698–4705.

Qian, Z., Ma, J., Shan, X., Shao, L., Zhou, J., Chen, J., et al. (2013). Surface functionalization of graphene quantum dots with small organic molecules from photoluminescence modulation to bioimaging applications: An experimental and theoretical investigation. *RSC Advances, 3,* 14571–14579.

Robinson, J. T., Tabakman, S. M., Liang, Y., Wang, H., Sanchez Casalongue, H., Vinh, D., et al. (2011). Ultrasmall reduced graphene oxide with high near-infrared absorbance for photothermal therapy. *Journal of the American Chemical Society, 133,* 6825–6831.

Shangguan, J., Huang, J., He, D., He, X., Wang, K., Ye, R., et al. (2017). Highly Fe(3 +)-selective fluorescent nanoprobe based on ultrabright N/P codoped carbon dots and its application in biological samples. *Analytical Chemistry, 89,* 7477–7484.

Shen, X., Liu, Y., Pang, Y., & Yao, W. (2013). Conjugation of graphene on Au surface by π–π interaction and click chemistry. *Electrochemistry Communications, 30,* 13–16.

Su, Y. H., Wu, Y. K., Tu, S. L., & Chang, S.-J. (2011). Electrostatic studies of π–π interaction for benzene stacking on a graphene layer. *Applied Physics Letters, 99,* 163102.

Suzuki, N., Wang, Y., Elvati, P., Qu, Z.-B., Kim, K., Jiang, S., et al. (2016). Chiral graphene quantum dots. *ACS Nano, 10,* 1744–1755.

Thakur, M., Kumawat, M. K., & Srivastava, R. (2017). Multifunctional graphene quantum dots for combined photothermal and photodynamic therapy coupled with cancer cell tracking applications. *RSC Advances, 7,* 5251–5261.

Wang, S., Cole, I. S., & Li, Q. (2016). The toxicity of graphene quantum dots. *RSC Advances, 6,* 89867–89878.

Wang, W., Li, Y., Cheng, L., Cao, Z., & Liu, W. (2014). Water-soluble and phosphorus-containing carbon dots with strong green fluorescence for cell labeling. *Journal of Materials Chemistry B, 2,* 46–48.

Wang, X., Cao, L., Yang, S.-T., Lu, F., Meziani, M. J., Tian, L., et al. (2010). Bandgap-like strong fluorescence in functionalized carbon nanoparticles. *Angewandte Chemie International Edition, 49,* 5310–5314.

Xu, X., Li, P., Zhang, L., Liu, X., Zhang, H.-L., Shi, Q., He, B., Zhang, W., Qu, Z., & Liu, P. (2017). Covalent functionalization of graphene by nucleophilic addition reaction: Synthesis and optical-limiting properties. *Chemistry—An Asian Journal, 12,* 2583–2590.

Xue, B., Zhu, J., Liu, N., & Li, Y. (2015). Facile functionalization of graphene oxide with ethylenediamine as a solid base catalyst for Knoevenagel condensation reaction. *Catalysis Communications, 64,* 105–109.

Yang, F., Lecroy, G. E., Wang, P., Liang, W., Chen, J., Fernando, K. A. S., et al. (2016). Functionalization of carbon nanoparticles and defunctionalization—toward structural and mechanistic elucidation of carbon "quantum" dots. *The Journal of Physical Chemistry C, 120,* 25604–25611.

Yang, S.-T., Wang, X., Wang, H., Lu, F., Luo, P. G., Cao, L., et al. (2009). Carbon dots as nontoxic and high-performance fluorescence imaging agents. *The Journal of Physical Chemistry C, 113,* 18110–18114.

Yoon, H. J., Kim, T. H., Zhang, Z., Azizi, E., Pham, T. M., Paoletti, C., et al. (2013). Sensitive capture of circulating tumour cells by functionalized graphene oxide nanosheets. *Nature Nanotechnology, 8,* 735–741.

Yuan, X., Liu, Z., Guo, Z., Ji, Y., Jin, M., & Wang, X. (2014). Cellular distribution and cytotoxicity of graphene quantum dots with different functional groups. *Nanoscale Research Letters, 9,* 108.

Zhang, L., Xia, J., Zhao, Q., Liu, L., & Zhang, Z. (2010). Functional graphene oxide as a nanocarrier for controlled loading and targeted delivery of mixed anticancer drugs. *Small, 6,* 537–544.

Zhang, L., Yu, J., Yang, M., Xie, Q., Peng, H., & Liu, Z. (2013). Janus graphene from asymmetric two-dimensional chemistry. *Nature Communications, 4,* 1443.

Zhang, Q., Jie, J., Diao, S., Shao, Z., Zhang, Q., Wang, L., et al. (2015). Solution-processed graphene quantum dot deep-UV photodetectors. *ACS Nano, 9,* 1561–1570.

Zhang, Y., Liu, C., Shi, W., Wang, Z., Dai, L., & Zhang, X. (2007). Direct measurements of the interaction between pyrene and graphite in aqueous media by single molecule force spectroscopy: Understanding the $\pi - \pi$ interactions. *Langmuir, 23,* 7911–7915.

Zhang, Z., Huang, H., Yang, X., & Zang, L. (2011). Tailoring electronic properties of graphene by π–π stacking with aromatic molecules. *The Journal of Physical Chemistry Letters, 2,* 2897–2905.

Zhao, G., & Zhu, H. (2020). Cation–π interactions in graphene-containing systems for water treatment and beyond. *Advanced Materials*, 1905756.

Zhu, Y., Higginbotham, A. L., & Tour, J. M. (2009). Covalent functionalization of surfactant-wrapped graphene nanoribbons. *Chemistry of Materials, 21,* 5284–5291.

Chapter 4
Graphene Nanomaterials for Multi-modal Bioimaging and Diagnosis of Cancer

4.1 Introduction

Carbon (C), the most abundant element found in nature, possesses different forms or allotropes. These allotropes can vary from three-dimensional candidates such as the carbon nanodiamonds (CNDs) all the way to the two-dimensional graphitic carbon (graphene or monolayer graphite), to one-dimensional carbon nanotubes (CNTs), and zero-dimensional fullerenes (Bartelmesse et al. 2015). With the breakthroughs and discoveries of the carbon allotrope—fullerene—by the Nobel laureates Sir Harold Kroto and Richard Smalley, carbon nanotechnology has fostered toward applications in the fields of physics, biomedical research, and energy applications. Figure 4.1 shows the most celebrated carbon nanomaterials in the field of nanotechnology.

With increasing carbon-based research, a new class of carbon nanomaterials is added to the group, and one of them is fluorescent graphene or carbon-based nanoparticles (CNPs). These fluorescent nanoparticles were accidentally discovered while purification of carbon nanotubes using electrophoresis (Fig. 4.2). Namely after that, many nomenclatures have been given following that decade till now as carbon nanotube dots (CNTDs), carbon quantum dots (CQDs), carbon dots (C-dots/CDs), and graphene quantum dots (GQDs) (Zhu et al. 2015). Furthermore, an another class of fluorescent carbon known as carbon nano-onions (CNOs) has been evaluated for many electrochemical applications (McDonough and Gogotsi 2013; Bartelmess et al. 2015).

R. Srivastava et al., *Next Generation Graphene Nanomaterials for Cancer Theranostic Applications*,
https://doi.org/10.1007/978-981-33-6303-8_4

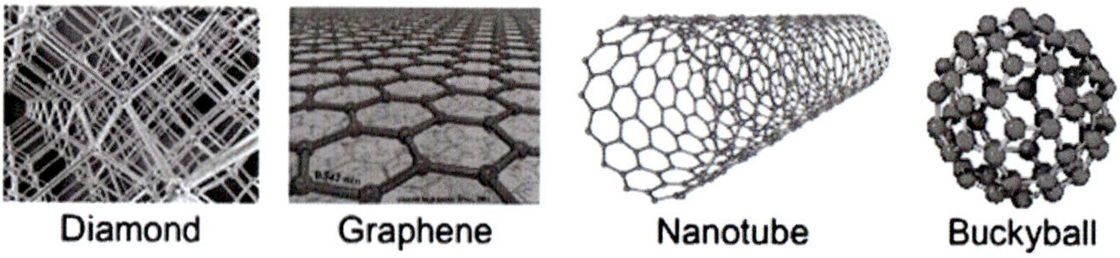

Fig. 4.1 Different dimensional yet stable forms of carbon most widely used in carbon nanotechnology applications. Adapted from Bartelmess et al. (2015)

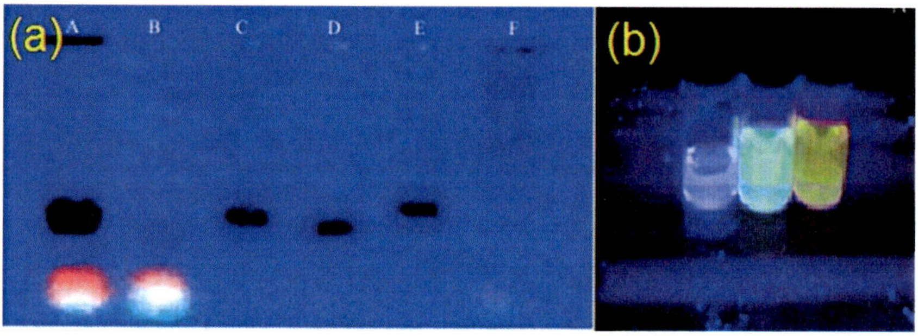

Fig. 4.2 **a** Electrophoretic analysis and UV illumination (365 nm) of carbon nanotube fragments giving the first direct evidence of fluorescent carbon, **b** extracted solution of the fluorescent carbon giving violet, blue, and yellow fluorescence. Adapted from Xu et al. (2004)

4.2 Origin of Fluorescence

The graphene-based materials—GO and GQDs (single-layer carbon core with connected chemical groups)—possess more defect, oxygen groups and functional groups on the surface. The excitons in graphene have an infinite Bohr diameter, thus showing quantum confinement effects as consequence of which the GQDs have a non-zero band gap and PL on excitation which is tunable by modifying the size and surface chemistry (Zhu et al. 2015). When the graphene sheets are cut along different crystallographic directions, diverse types of edges (armchair and zigzag edges) are obtained. The edge type plays an important role in determining the electronic, magnetic, and optical properties. The PL mechanism can be attributed to the surface/edge state and conjugated π-domains. The surface/edge state possesses triple carbene at the zigzag edges, oxygen-based groups on the graphene core, and resonance of amine moieties and graphene core. Blue PL of hydrothermally GQDs might be attributed to free zigzag sites with a carbine-like triplet ground state described as $\sigma 1$-$\pi 1$ (Pan et al. 2010) (Figs. 4.3 and 4.4).

Photo-excited electrons through the π-π^* transitions relax into either sp^2 energy levels or the defect states (surface state), giving rise to the blue or long-wavelength PL, respectively. The former emission might bear the discrete feature due to

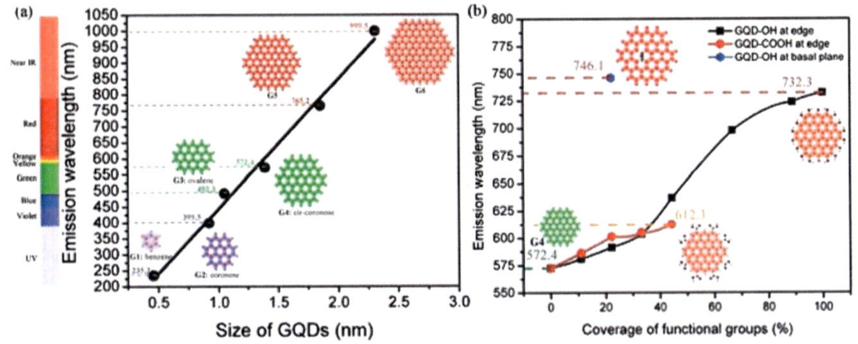

Fig. 4.3 Emission wavelength as a function of size and functional groups (–OH, –COOH) on the oxidized graphene quantum dots. The edged sites in are zigzag configurations. The study is adapted from Mahasin Alam et al. (2014) and Zheng et al. (2015)

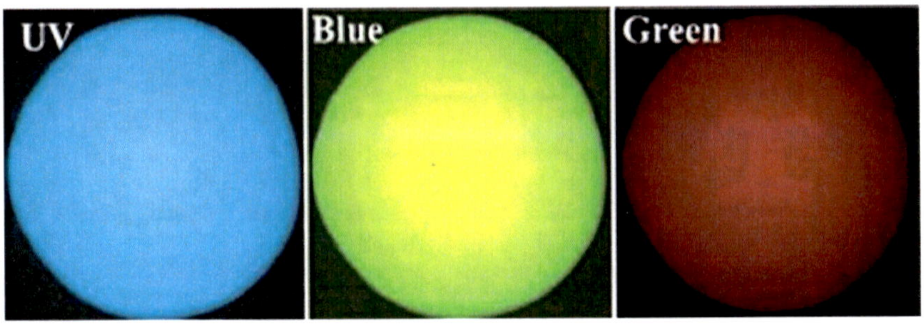

Fig. 4.4 Photoluminescence photographs of carbon dots in water under ultraviolet (330–385 nm), blue (450–480 nm), and green (510–550 nm) light excitation. Adapted from Li et al. (2012)

quantum confinement effect of electrons inside the sp^2 carbon domains. The latter emission is related to the hybrid structure by both the oxygen functional groups (at the edges and/or on the basal planes) and graphene core. Despite noticeable differences in the size and the number of layers from particle to particle, all of the GQDs studied possess almost the same spectral line shapes and peak positions. As a result, it suggested the PL of these GQDs was caused by surface state. The fluorescence emission arises from both an intrinsic bandgap resulting from confined *sp*2 conjugation in the core of the GQDs and extrinsic fluorescence resulting from surface state that can be either directly excited or excited by energy transfer from intrinsic band (Wang et al. 2010). As a result, the tunable fluorescence emission of CQDs can be achieved by either controlling the domain size of *sp*2 conjugation or modifying the chemical groups formed on the surface of CQDs (Hou et al. 2013 and Oza et al. 2015).

Graphene-based nanomaterials are structurally abundant in C, H, O and other elements such as S based on the synthesis strategy. Therefore, it is considerable to

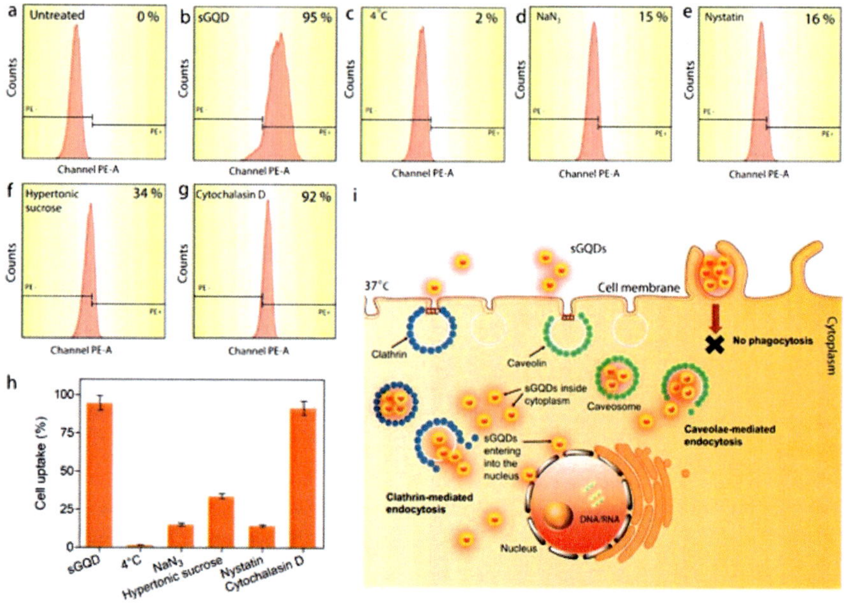

Fig. 4.5 Mechanism of cellular intake of graphene-based quantum dots nanomaterial ('sGQDs') as shown by flow cytometry analysis. In comparison with non-treated cells (**a**), the treated cells (**b**) show higher fluorescence correlating with higher cellular uptake (>90%). Adapted from Kumawat, Thakur, et al. (2017)

believe that these nanomaterials may high low toxicity and thereby minimum side effects when ingested into the body. Although most of the reported biocompatibility of these nanomaterials is usually high, based on the precursors of synthesis and exposed surface groups, these graphene-based nanomaterials may induce deleterious effects inside the cell. Hence, it is important to discuss fundamental question like: How these nanomaterials react with mammalian cells? Table 4.1 describes some of the important pitfalls and possible solutions these nanomaterials undergo when they interact with cells. Graphene-based nanomaterials such as GO (functionalized) and other nanomaterials including GQDS, CDs, and CNOs have advantageous properties such as wide range of optical absorption and fluorescence emission property (imaging modality), biocompatibility, and biodispersibility in many solvents including biofluids such as blood and serum. Furthermore, the size and charge on these nanomaterials are inherently neutral toward any cellular response and can also be modified (or functionalized) for better cell biocompatibility and open a possibility to attach additional biomolecules or drugs. These conjugation strategies along with inherent optically active fluorescence properties make an attractive tool for therapy and imaging, often referred to as 'theranostics' (Fig. 4.6).

Table 4.1 Physiological effects of graphene-based nanomaterials in cells relevant to imaging and therapeutic applications

Biorelevant properties	Issue(s)	Solution(s)	References
Aggregation	Aggregation in cell media and in different ionic solutions	Oxidation or functionalization for increasing hydrophilicity	Zheng et al. (2015), Li et al. (2017) and Yan et al. (2019)
Dispersibility	Similar to aggregation due to hydrophobic residues	Functionalization of the exposed residues with polar residues	Cioffi et al. (2009), Luo et al. (2013), Pandey, Mewada, et al. (2013), Tripathi et al. (2013), Pandey et al. (2014), Thakur et al. (2014, 2016), Zhao et al. (2015), Kumawat, Thakur, et al. (2017) and Tian et al. (2018)
Cellular uptake (also Fig. 4.5)	Size-dependent uptake and adsorption on the cell wall	Lateral dimensions hinders cell uptake; hence, the nanomaterials need to be in the order of endosomes or even smaller	Mu et al. (2012), Chatterjee et al. (2014) and Zhang et al. (2016)
Biocompatibility	Concentration-dependent effect on cells	Higher concentration causes destabilization of cell membrane, reactive oxygen species production, etc.	Li et al. (2012), Puvvada et al. (2012), Sonkar et al., (2012), Luo et al. (2013), Thakur et al. (2014), Kumawat, Srivastava, et al. (2017), Kumawat, Thakur, et al. (2017) and Tian et al. (2018)

4.3 Light-Based Imaging

One of the most promising and very often emphasized applications of fluorescent graphene-based materials is in biomedical imaging. These fluorescent materials offer a useful platform for exploring the challenges in the field of nanomedicine, in particular in diagnosis and in cancer therapy. The large surface area of these nanomaterials and the availability of carboxylic functional groups on the surface

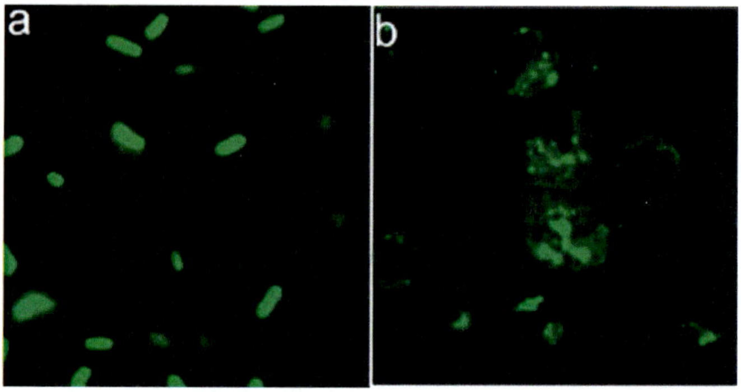

Fig. 4.6 Confocal fluorescence microscopy images (488 nm excitation) of **a** *E. coli* and **b** Caco-2 cells labeled with CQDs. Adapted from Luo et al. (2013)

allow for preparation of multi-modal probes and therapeutic conjugates. Major properties of these fluorescent materials include exceptional biocompatibility, high aqueous/lipophilic solubility, excitation-dependent fluorescence, wide excitation with narrow emission spectrum, non-blinking florescence, long fluorescent lifetime, and resistant to photobleaching (Li et al. 2012). They surpass traditional semi-conductor quantum dots with cheap precursors and biocompatibility (Puvvada et al. 2012; Zhou et al. 2012; Luo et al. 2013). Various classes of fluorescent CNPs have been discussed in the literature (Krueger 2008; Puvvada et al. 2012; Sonkar et al. 2012; McDonough and Gogotsi 2013; Tripathi et al. 2013; Ye et al. 2013; Du et al. 2014) (Fig. 4.7).

4.3.1 *In Vitro Bioimaging*

Due to biocompatibility and aqueous solubility, fluorescent CQDs have been primarily used for biolabeling applications. Since a decade after their discovery, CQDs have been used to image a simplest bacterium to mammalian cells in vitro (Cioffi et al. 2009; Pandey, Mewada, et al. 2013; Pandey, Thakur, et al. 2013; Oza et al. 2015). Mostly, cells incubated with CQDs are biolabeled either by cells taking nanoparticles inside them or adsorption of CQDs onto cell membrane. Functional group of CQDs can affect its cellular internalization (H. Sun et al. 2013). Sharon and co-workers have used CQDs for optical imaging of *Saccharomyces cerevisiae*, mammalian cell lines such as Vero, HeLa, and MCF-7 in vitro (Pandey, Thakur, et al. 2013; Thakur et al. 2014) (Fig. 4.8).

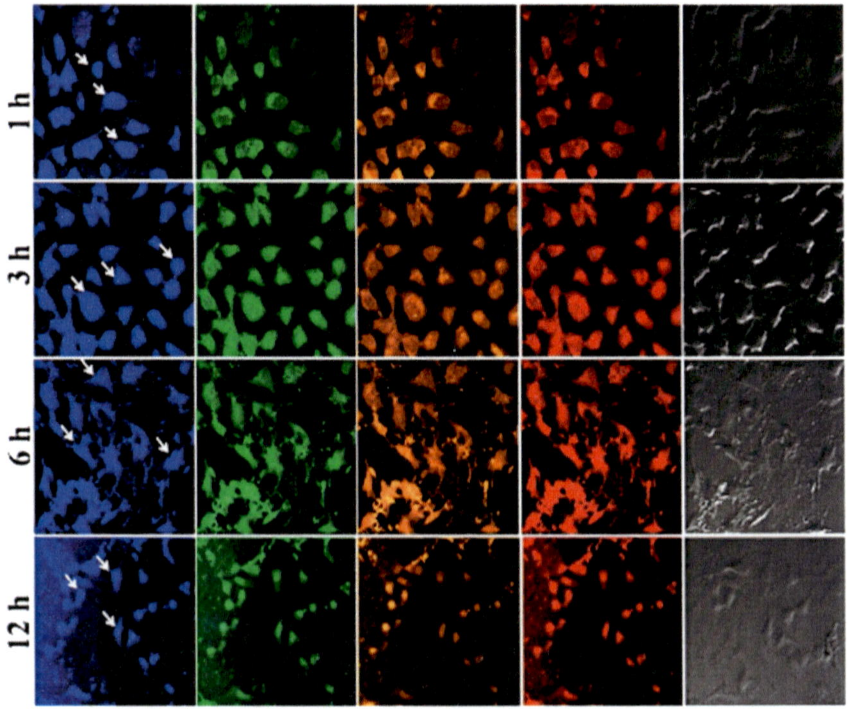

Fig. 4.7 Confocal laser scanning microscopic CLSM) images of MDA-MB-468 cells treated with CQDs after 1, 3, 6 and 12 h. Excitation and emission wavelengths: first column: λex = 359 nm, λem = 461 nm; second column: λex = 494 nm, λem = 520 nm; fourth column: λex = 570 nm, λem = 590 nm; the third column is a sum of the second and fourth columns. Scale bars: 1, 3, 6 h–10 μm, and 12 h–20 μm. Adapted from Puvvada et al. (2012)

4.3.2 *In Vivo Bioimaging*

Owing to their low or minimal cytotoxicity, CQDs have been used for in vivo bioimaging applications. Tao et al. used nude mice model to image with an aqueous solution of CDs which was injected subcutaneously into the mice, followed by fluorescence imaging with excitations at seven different wavelengths from 455 to 704 nm (Tao et al. 2012). In one study, PEGylated carbon dots were injected fluorescence images at 470 nm excitation were collected continuously. Following the injection, carbon dots migrated along the arm to the axillary lymph node. However, the observed migration was slower in comparison with that of the semiconductor QDs (Luo et al. 2013) (Fig. 4.9).

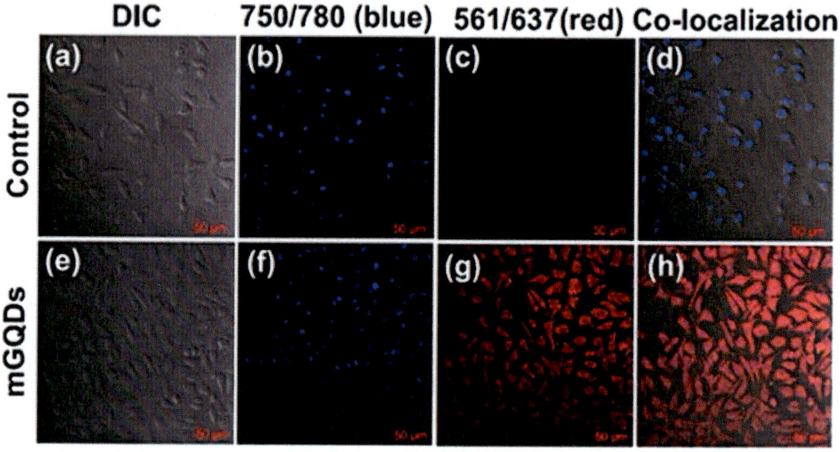

Fig. 4.8 Confocal laser scanning microscopic (CLSM) images of NIR fluorescence imaging of L929 cells from GQDs made from Mango leaves. Adapted from Kumawat et al. (2017)

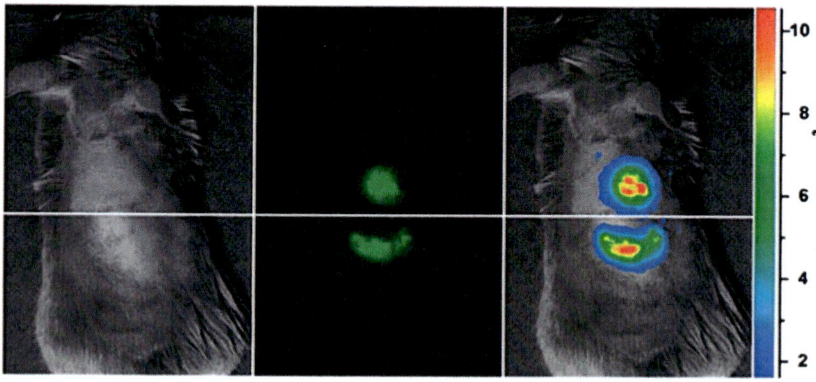

Fig. 4.9 A comparison of in vivo imaging results (subcutaneous injection on the back of mice) between the carbon dots (top) and commercial CdSe–ZnS QDs (Invitrogen, aqueous Qdot-525, bottom), with the equal number of dots under the same imaging conditions (434 nm band-pass for excitation and 474 nm cutoff filter for emission). Adapted from Luo et al. (2013)

4.4 Contrast-Based Imaging

4.4.1 X-Ray Imaging

Computed tomography (CT) scans are advanced imaging techniques wherein X-ray is used in combination with computer-aided software to give cross-sectional images inside a patient, allowing to see inside the user without cutting. Tomography is derived from two words 'tomos' which means slices and 'graphein' which translates to 'to write.' Tomography is defined as imaging of an object by analyzing its slices. It is also known as computer axial tomography (CAT). Medical imaging is the most common application of X-ray CT. The basic purpose in any radiologic examination is to recognize abnormalities while reducing the radiation exposure (Huda 2014b).

X-rays are generated when electrons from the cathode accelerated by high voltage collide with a metal anode. In CT, tungsten is one of the most common anode materials due to its high atomic number and high melting point. Bremsstrahlung and characteristic radiation are two processes that are responsible for producing X-rays (Lee et al. 2013) (Fig. 4.10).

The first-generation CT machines had an X-ray tube and single detector which moves together by translation and rotation, whereas the X-ray beam had a linear shape. The second-generation CT advanced to multiple detectors arranged in one row having similar motion as their first-generation counterparts. The X-ray beam

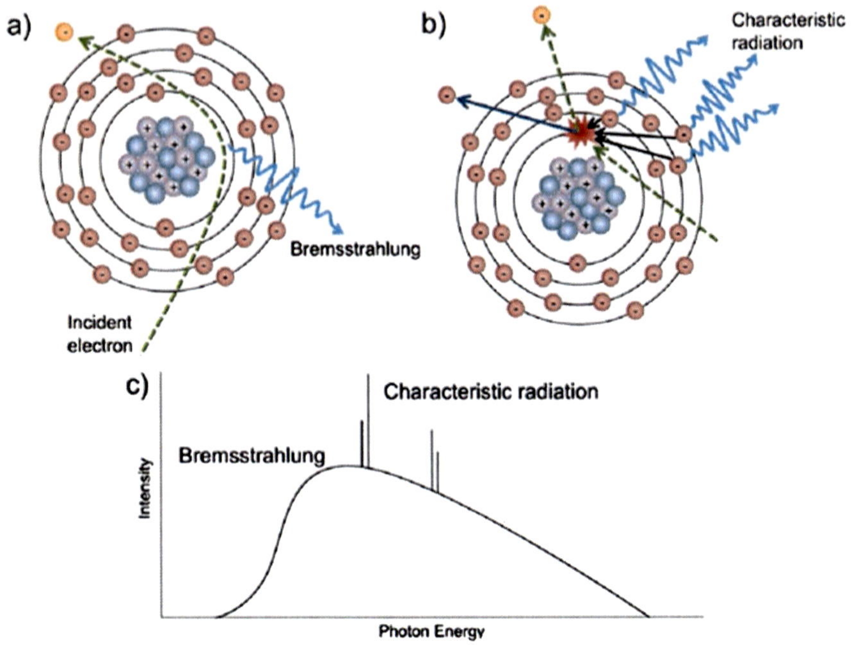

Fig. 4.10 **a** Incident electrons are scattered due to the electric field of the anode nuclei, X-ray photons with continuous spectrum are generated (also known as Bremsstrahlung radiation). **b** Incident electron with higher energy than the *K*-shell binding energy collides with electrons in the inner shell, resulting in ejection of the bound electron which in turn leads to a vacancy creation. The vacancy when filled with an outer electron leads to emission of X-ray having energy equal to the difference in binding energies of the outer shell and *K*-shell (also known as characteristic radiation). **c** Typical X-ray emission spectrum showing Bremsstrahlung and characteristic radiation. Adapted from Lee et al. (2013)

had a fan-like shape. Further, the third-generation CT scanners had a system of completely rotating X-ray tube and detectors complex; however, the rotating detector was dropped in the fourth generation. Currently, the fifth generation is being utilized wherein patient is rotated on the table and slip ring technology permits transmission of energy to rotating gantry without requiring cables (Goldman 2007).

Several issues plague CT scan technique such as high radiation dose, exorbitantly expensive, laziness among clinicians, among others. It should be advised to all CT scan practitioners to advise CT scanning only when it is utmost required and reduce the radiation dose to the least level especially the ones dealing with children (Fred 2004). Increasing X-ray photons can reduce the quantum mottle, which causes noise in all radiographic imaging. This, however, will lead to an increase in the patient dose (Huda 2014a). Image contrast is majorly determined by average photon contrast, which is controlled by the amount of X-ray beam filtration and X-ray tube voltage. Contrast and noise are directly related to the choice of radiographic techniques, namely X-ray tube voltage (kilovoltage) and output (tube current–exposure time product). The choice of tube current (milliamperes) and X-ray beam exposure time (seconds). Air kerma is the kinetic energy released per unit mass when X-ray interacts with air, which is used to quantify the X-ray intensity (Frush and Ogden 2008) (referenced in 4). The quality of the X-ray is generally defined by its penetrating power.

The inherent sensitivity of CT is not sufficient; therefore, contrast agents are often needed to detect subtle change of soft tissues. A good CT contrast agent needs a high density of heavy atoms having high X-ray absorption cross sections which helps in creating contrast forming a shadow to the X-ray illumination. X-ray attenuation effect of a material usually increases with its atomic number; therefore, high-z elements are preferred CT contrast agents. (Sandborg et al. 1995) Iodine-containing vesicles or micelles, lanthanide, and gold nanoparticles are the most common contrast agents for X-ray CT imaging due to availability of heavy atoms in their chemical compositions. Contrast agents need to enter and pass through anatomical regions to provide contrast enhancement. There are two types of contrast agents at present approved for human use, $BaSO_4$ suspensions which can only be used for gastrointestinal (GI) tract imaging, due to the toxicity induced by Ba^{2+} ions. The second type of contrast agent is water-soluble aromatic iodinated contrast agent (Yu and Watson 1999). Several experimental and theoretical determinations have shown heavy metal elements to be good contrast agents (Sandborg et al. 1995; Kim et al. 2018).

Another factor that needs to be taken care of for a good CT agent is the cost, because a high dose is usually required for a single scan. Therefore, even though gold-based CT agent is an attractive option because of its facile synthesis and good biocompatibility, the cost makes it unusable for all practical purposes. It is predicted that a good contrast agent will be useful in reducing both the dose and radiation exposure (Fig. 4.11).

Carbon nanomaterials such as graphene oxide do not have good contrast properties; however, they can be made suitable for CT agent by conjugating it with

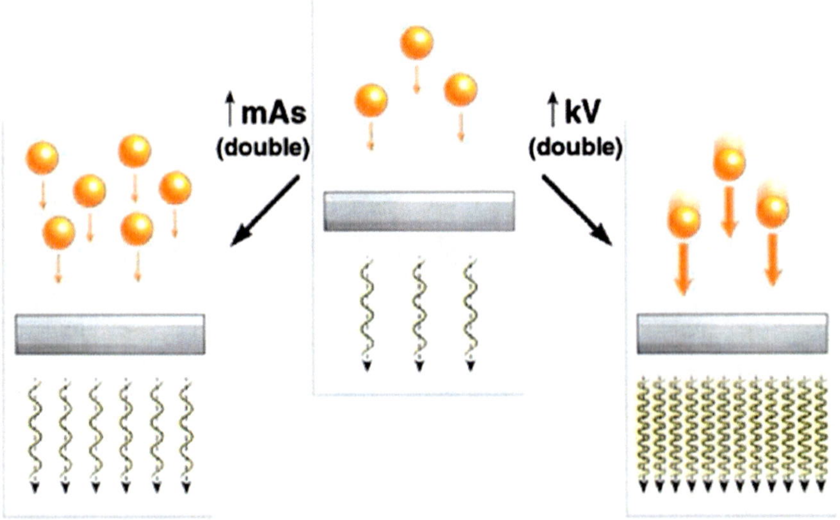

Fig. 4.11 A number of X-rays produced increases with increase in milliampere-second (mA s) value with no change in average photon energy (left). On the right, we can see that with an increase in tube voltage, more photons are produced which increases the average photon energy. Increasing mA s only increases quantity, whereas with kV, we can see an increase in both quality and quantity. Adapted from Huda and Abrahams (2015)

high-*z* molecules such as gold, barium, and iodine while utilizing GO's absorbance in the NIR region for photothermal application. Zhou et al. (2014) used graphene/Ni nanoplatelet for X-ray chemical imaging to analyze their local electronic and chemical structure. Sun et al. (2017) synthesized graphene oxide conjugated with gold nanorods (GO/GNRs) by gold-seed in situ growth method at room temperature. The GO/GNRs were injected into xenograft tumors; excellent computed tomography (CT) imaging properties and photothermal effect were obtained. Zhang et al. (2015) synthesized GO/$BaGdF_5$/PEG which enabled dual-modality MR and X-ray imaging owing to the presence of Gd^{3+} and Ba^{2+} in the complex, apart from having exceptional NIR absorbance, photothermal stability, and effective tumor passive targeting. The GO/$BaGdF_5$/PEG was also used in vivo for MR imaging, CT imaging and photothermal therapy. Shi et al. (2014) used chemical deposition of Ag nanoparticles onto GO to form a GO@Ag nanocomposite through a hydrothermal reaction. Doxorubicin, an effective cancer drug, was linked to it via ester bond. The GO@Ag-Dox complex served as an effective X-ray contrast agent and also as strong agent for photothermal agent. Another group, Shi et al. (2013) conjugated iron nanoparticles and gold to graphene oxide and coated the compound with PEG to form GO-IONP-Au-PEG which utilized the magnetic properties of iron for MR scanning and high-Z property of Au for effective CT imaging. The absorbance of GO in NIR region also made it an effective photothermal agent, thereby making it useful for photothermal therapy apart from its usage for bimodal imaging. This was

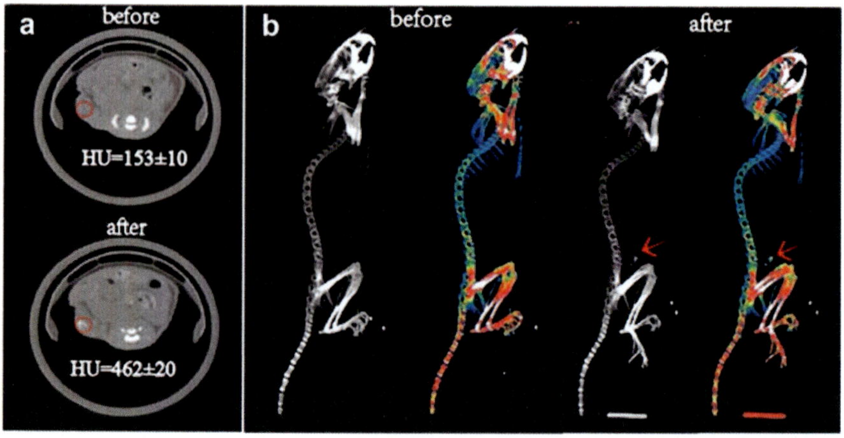

Fig. 4.12 CT imaging of mice in vivo. **a** X-ray transverse CT images of tumor before and after intratumoral injection. The CT values of tumor site are 153 ± 10 HU (before injection) and 462 ± 20 HU (0.5 h after injection); **b** the 3D VR CT images before and after intratumoral injection. Adapted from Zhang et al. (2015)

also examined in vivo by the group. Zhang et al. (2017) used a one-pot solvothermal method to prepare GO/Bi_2Se_3/PVP nanocomposites by direct deposition of Bi_2Se_3 NPs on GO in the presence of polyvinyl pyridine. These nanocomposites can be used as good computed tomography/photoacoustic dual-modality imaging systems due to the strong NIR optical absorbance and X-ray absorption coefficient. Lalwani et al. (2014) synthesized graphene nanoplatelets with Mn intercalation and further functionalized with iodine, which can be used as bimodal contrast imaging agent for MRI and CT due to the presence of Mn which helps in MRI scanning, and iodine functionalization is useful for CT scans.

All the bimodal contrast agents such as the ones developed by Lalwani et al., Zhange et al., and Shi et al. have a good scope of translating into the first clinically approved bimodal contrast agent. The properties of the standard contrast agents functionalized/attached with graphene oxide can be a good coupled system for biomedical imaging and photothermal therapy (Fig. 4.12).

4.4.2 *Magnetic Resonance Imaging*

Magnetic resonance imaging (MRI) technique is based on the magnetization properties of atomic nuclei. MRI is a medical imaging technique that is used in radiology forming pictures of the anatomy and the processes in the body in good health as well as diseases. The randomly oriented protons present in the tissue water are aligned by an external uniform magnetic field. The aligned is disordered after introducing external radio frequency (RF) energy. Fourier transformation is used to convert the frequency information in the signal from each location in the imaged plane to corresponding intensity levels, which are then displayed as shades of gray in a matrix arrangement of pixels. This alignment (or magnetization) is net perturbed or disrupted by introducing external radio frequency (RF) energy. Following the initial RF, the emitted signals are measured. Fourier transformation is used to convert the frequency information contained in the signal from each location in the imaged plane to corresponding intensity levels, which are then displayed as shades of gray in a matrix arrangement of pixels. By varying the sequence of RF pulses applied and collected, different types of images are created. Repetition time is the time between successive pulse sequences.

Tissue can be characterized by two different relaxation times—T_1 and T_2. T_1 (longitudinal relaxation time) is the time constant which determines the rate at which excited protons return to equilibrium. It is a measure of the time taken for spinning protons to realign with the external magnetic field. T_2 (transverse relaxation time) is the time constant which determines the rate at which excited protons reach equilibrium or go out of phase with each other. It is a measure of the time taken for spinning protons to lose phase coherence among the nuclei spinning perpendicular to the main field.

MRI does not contain X-rays, which distinguishes it from CT or CAT scans. Its non-invasiveness coupled with spatial detail finer than a millimeter makes it a very attractive contrast imaging technique. The molecule which causes most functional brain imaging is an example of a good MRI contrast imaging agent. The MRI method is based on the magnetic property of protons that align themselves in a very large magnetic field, and the main MRI parameters are spin-lattice (T_1) and spin–spin (T_2) relaxation rates and times. The contrast agents (CAs) are used to shorten relaxation times and can be classified in several ways based on their chemical nature, magnetic properties, biodistribution, etc. (Jasonoff 2006). Most of the CAs, including the most widely used gadolinium-based CAs, modify T_1 relaxation. However, the toxicity of these Gd-based CAs is being investigated. MRI CAs based on the T_2 exchange mechanism have more recently expanded the armamentarium of agents for molecular imaging. Compared with T_1 and T_2* agents, T_2 exchange agents have a slower chemical exchange rate, which improves the ability to design these MRI CAs with greater specificity for detecting the intended biomarker (Sinharay and Pagel 2016).

Variety of optical imaging techniques are used for their source of contrast such as fluorescence, absorbance, reflectance, and bioluminescence. Near-infrared probes are highly beneficial in place of lower-wavelength excitation light because NIR light can penetrate deeply into tissues to generate emission from probe molecules unlike lower-wavelength light (Padalkar and Pleshko 2015). The need for new CAs in MRI is the need of the hour, but this is limited due to access of the advanced MRI scanners and huge electronic equipment for the potential of the molecular agents to be realized. Increasing availability of these sensitive detection methods, along with high field magnets and a growing collection of smart CAs, promises to potentiate MRI as an important tool for fundamental biology research.

Contrast agents are divided into two categories. The first one is paramagnetic compounds, including lanthanides like gadolinium, which mainly reduce the longitudinal (T_1) relaxation property and result in a brighter signal. The second class consists of superparamagnetic magnetic nanoparticles (SPMNPs) such as iron oxides, which have a strong effect on the transversal (T_2) relaxation properties. Magnetic fluids have high r_2/r_1, especially at high field strength of 1.5*T*.

Numerous techniques and technological advancements are being developed to improve the contrast agent's properties and increase the image contrast. One of the techniques to increase the imaging contrast is to combine it with other highly sensitive systems like proton emission tomography (PET), smart probes or agents which leads to very high *S*/*N* ratio and increased sensitivity. On particle size effect, Brooks et al. (2001) showed that for increasing particle size of single-core size MNPs, the magnetization and r_2 relaxivity increase. On the other hand, Xie et al. (2018) showed that size effect becomes less significant for the single-core iron oxide nanoparticles above 50 nm. Therefore, magnetic nanoclusters are an effective way to enhance the sensitivity due to their small size and increased effective magnetic size as shown by Chen et al. (2010) MR contrast agents shorten T_1 and T_2, which satisfies several requirements for clinical applications such as adequate relaxivity and susceptibility effects, tolerance, safety, low toxicity, stability, optimal biodistribution, and elimination (Shokrollahi 2013).

Gadolinium-based contrast agents are widely used as MRI-based contrast agents because in the 3+ oxidation state the metal has seven unpaired electrons. This causes water around the contrast agent to relax quickly, enhancing the quality of the MRI scan. The Gd^{3+} is paramagnetic in nature. In the late 1990s, Prince et al. (Frisoli and Ph, no date) reported that Gd-based contrast agents were far less nephrotoxic than their iodinated counterparts. The connection between nephrogenic systemic fibrosis and gadolinium-based contrast agents was suggested later after six years (Thomsen 2007; Caravan et al. 1999).

Ever since its discovery in 1991 (Iijima 1991), carbon nanotubes (CNTs) show unique physical properties (Eatemadi et al. 2014) and can be produced at high yields (Ebbesen and Ajayan 1992). The small size, robustness, and exceptional physical properties make them suitable for a host of various applications (Ajayan and Zhou 2001; Ajayan et al. 2000). Their applicability in biomedical applications (He et al. 2013) as a therapeutic and diagnostic tool and the abundance of functional groups on its surface to carry imaging agents makes it a suitable candidate as a contrast agent. The unique and exceptional properties of Gd were used further by functionalizing CNT with Gd to use it as MRI cell labeling and tracking. For MRI applications, the coupling of CNTs with iron oxide superparamagnetic nanoparticles has also been explored for developing contrast agents with effect on T_2 and T_2* relaxation times (Servant et al. 2016).

The Gd-functionalized carbon nanotubes also known as gadonanotubes (GNTs) are up to 100 times more effective than current clinical CAs which is a key in attaining both cellular and molecular imaging using MRI (Sitharaman and Wilson 2006). The carbon nanotubes can be functionalized non-covalently and covalently (Servant et al. 2016; Chen et al. 2003). Another technique commonly employed is forming heterostructured complex between magnetic iron oxide nanoparticles and CNTs. These help by decreasing the spin–spin relaxation time T_2 and creating image contrast while retaining CNTs optical properties. Magnetic nanoparticles have higher relaxivity than Gd(III)-chelating molecules, which are used for spin–lattice relaxation time T_1—weighted imaging. These properties of FeNPs make them a suitable tool as mediators for magnetic hyperthermia and MRI CAs (Mornet et al. 2004).

The use of carbon nanomaterials has sEmission wavelength as a function een recent interest in the development of contrast agents for MRI. Chaudhary et al. (2017) synthesized Fe core–C shell nanoparticles which can achieve substantially higher relaxivity of 70 mM^{-1} S^{-1}. Fullerenes have been functionalized with Gd to form metallofulerene complexes which showed potential application in MRI. The metal ion gets entrapped in the fullerene interior space, due to the superchelating nature of fullerene. The medical applications of fullerene can be employed in conjugation with the Gd- or iron-based contrast agent (Kato et al. 2003; Bolskar et al. 2003; Sun et al. 2015). Further, carbon nanostructures such as nanodiamonds have also been functionalized using Gd and magnetic nanoparticles to be used as fluorescent agents (Lien et al. 2012; Chang et al. 2008), contrast enhancement agents (Manus et al. 2010), inducing ferromagnetism using trivalent impurities (Talapatra et al. 2005).

The use of carbon nanostructures further generated interest among scientists to use graphene for MRI while utilizing its various biomedical applications, especially photothermal therapy in case of MRI for live, real-time imaging of in vivo agent distribution and post-treatment therapeutic outcomes. Cong et al. (2010) showed tunable magnetism by controlling the coverage density of magnetite on GO sheets and opened up possibilities to use graphene as MR imaging agents. Zhang et al. (2018) developed a pH-responsive nanosystem containing GO with folic acid and Gd-labeled dendrimer to boost its T_1 contrast ability further loaded with doxorubicin and colchicine. This system showed an ultrahigh relaxivity of up to 11.6 mM^{-1} s^{-1} and excellent magnetic resonance angiography (MRA) images with high-resolution vascular structures because of long blood circulation time of the folic acid–gadolinium-labeled dendrimer.

Chen et al. (2011) used aminodextran coated with iron oxide nanoparticles and graphene oxide using EDC chemistry for T_2 MRI contrast imaging, wherein the complex showed better T_2 weighted MRI contrast due to formation of aggregates of Fe_3O_4 on GO enhancing T_2 relaxivity. Yang et al. (2012) employed a similar strategy by using graphene nanosheets anchored with magnetic nanoparticles for use in photothermal therapy along with multi-modal imaging. Ma et al. (2012) fabricated GO-IONP-PEG for magnetically targeted drug delivery, photothermal therapy serving as one system for multiple applications of graphene oxide-modified compounds for MR-based applications. Another similar system was developed by Zhang et al. (2013) for MRI and drug delivery using GO-Gd complex. The wide use of graphene has interested the researchers to further use other 2D materials for MRI (Pan et al. 2018).

4.4.3 Radionuclide-Based Imaging

Positron emission tomography (PET) is a highly sensitive non-invasive technology that is ideally used for imaging of cancer biology in both preclinical and clinical stages. Positron emission tomography (PET) imaging is based on positron-emitting radionuclides being visualized and quantified. This technique gives high sensitivity and accurate quantifiable estimations of the radiotracer due to the excellent tissue-penetrating properties of photons. The positron emitted generates two 511 keV photons after annihilating nearby electrons which are detected by detectors embedded in PET scanners. Examples of routinely used positron-emitting isotopes are ^{11}C, ^{13}N, ^{15}O, 18F, ^{44}Sc, ^{62}Cu, ^{64}Cu, ^{68}Ga, ^{72}As, ^{74}As, ^{76}Br, ^{82}Rb, ^{86}Y, ^{89}Zr, and ^{124}L[26–33]. Single-photon emission tomography (SPECT) is another similar which unlike PET utilizes non-coincident γ-rays generated by radionuclides. SPECT suffers from few disadvantages compared to PET such as lack of anatomical information and low spatial resolution. However, some of the disadvantages are overcome by utilizing hybrid techniques such as SPECT combined with CT.

The surface chemistry of GO which is easily tunable greatly affects its toxicity. Chen et al. (2017) functionalized GO with PEG for enhancing amino groups using covalent conjugation. Graphene oxide has also proved to be very efficient in removal of radionuclides due to its high surface area and multiple surface functionalities. Y. Sun et al. (2013) chemically grafted polyaniline on graphene oxide composite for efficient removal of radionuclides in a wide range of acidic to alkaline medium. They studied the interaction mechanism between radionuclides and GO-PANI by using XPS and concluded that the sorption occurs by forming complexes with oxygen- and nitrogen-containing functional groups. Further, it was observed that the radionuclides have a higher affinity to nitrogen-containing functional groups as compared to oxygen-containing functional groups.

4.5 Raman Spectra-Based Imaging

Raman spectroscopy has become one of the most diverse analytical tools in biomedical as well as chemical research. C. V. Raman was given the Nobel Prize for discovering the Raman effect in 1928 which talks about the inelastic scattering of photons by molecules which are excited to a higher vibrational or rotational energy level. There are two types of Raman scattering where the energy exchange occurs between the incident photon and a molecule: Stokes scattering (molecule absorbs energy) and anti-Stokes scattering (molecule loses energy). The unique energy level and frequency of light scattered being unique for all the molecules make Raman a highly chemical-specific technique.

Over the years, with the improvement in lasers and instrument design, the sensitivity of Raman has increased and the fluorescence interference is not a huge

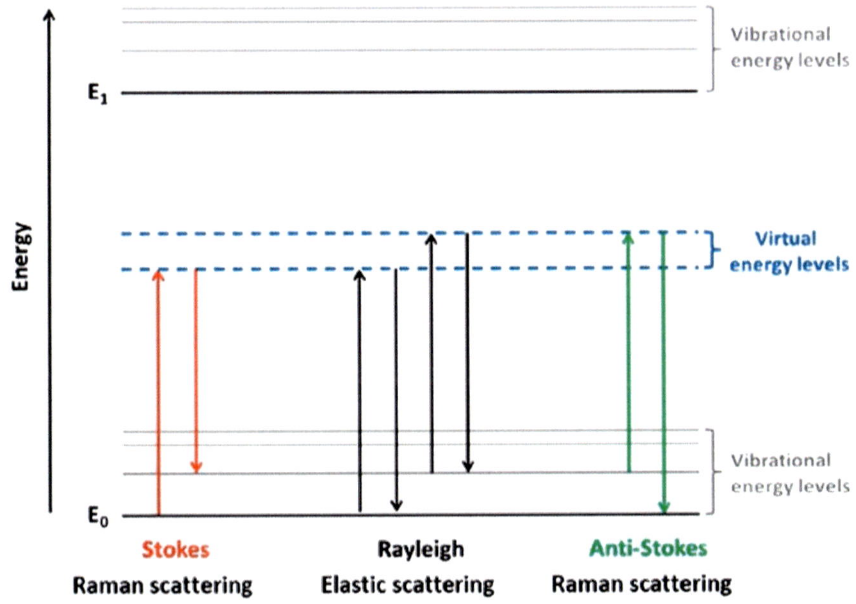

Fig. 4.13 Jablonski style diagram explaining different types of energy transitions involved wherein Stokes scattering, the incident photon is of greater energy the scattered photon and in anti-Stokes scattering, and the incident photon is of lower energy. Adapted from Ember et al. (2017)

challenge owing to the improved and diversified lasers along with sensitive detectors. In modern times, several techniques have been developed for Raman spectroscopy-based imaging such as coherent anti-Stokes Raman spectroscopy (CARS) and surface-enhanced Raman scattering (SERS). Raman has been used for diagnosis of various kinds of cancer such as lung, (McGregor et al. 2017) breast, and (Haka et al. 2005, 2006) bladder (Kerr et al. 2016; Li et al. 2015) (Fig. 4.13).

Delhaye and Dhamelincourt (1975) showed in 1975 the possibility of using Raman as a microscope and a microprobe by adding spatial resolution. Raman spectroscopy makes an ideal tool for diagnosis and characterization because Raman spectrum is like a fingerprint, wherein every chemical will have a unique spectrum. Coherent anti-Stokes scattering also known as CARS is a process which involves three laser fields at the pump, Stokes, and probe frequencies interact with the medium to generate a new field at the anti-Stokes frequency (Fig. 4.14).

The 2D band which defines the stacking order is a facile tool to study the presence of single-layer graphene as it originates from a two-phonon double-resonance process. Even though AFM is the most direct way of measuring number of graphene layers, it has its own limitations such as low throughput and an instrumental offset of 0.5 nm of monolayer. Ferrari et al. (2006) studied how the D' line can be used to distinguish single-layer graphene and double and few layers. Lee et al. (2017) showed Raman as a powerful tool to study the nature of

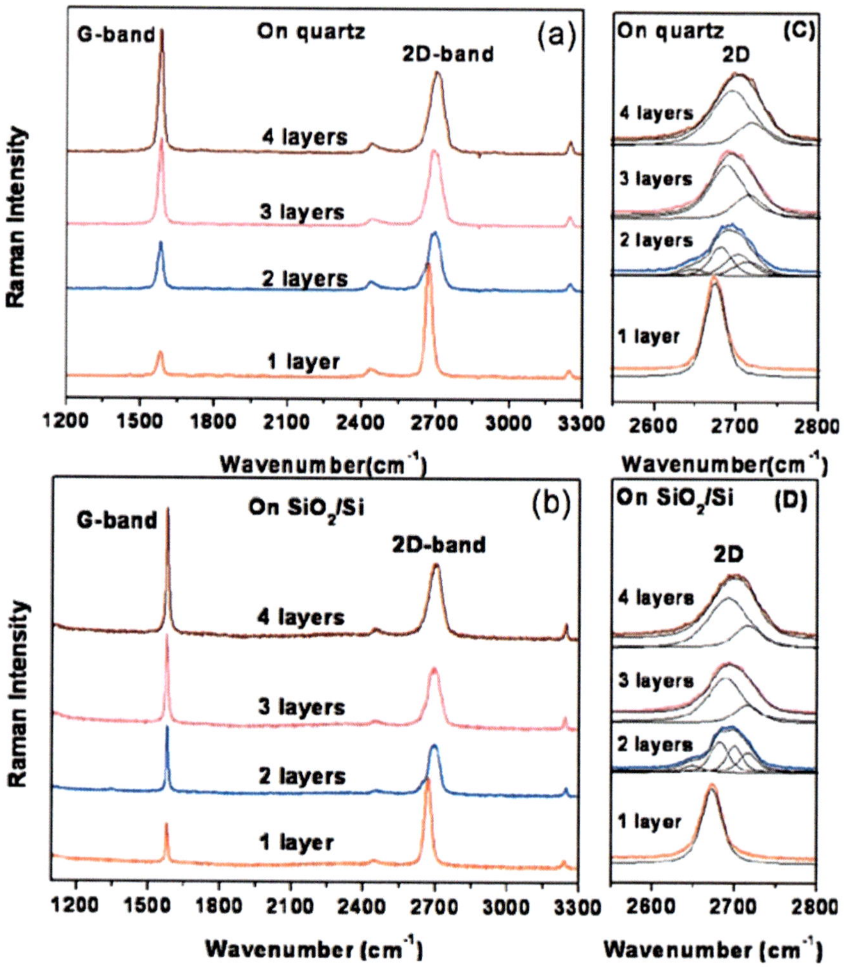

Fig. 4.14 **a**, **b** Raman spectra of monolayer, bilayer, three layers, and four layers on graphene and SiO_2 (300 nm/Si substrate). **c**, **d** The enlarged 2D bands with curve fit. Adapted from Wang et al. (2008)

defects in graphene. The team observed that the nature of defects varies greatly based on whether it is chemically or anodically derived and further corroborated the results with AFM results. Tip-enhanced Raman spectroscopy (TERS) is also studied as a powerful nanoanalytical tool to study the topographical and molecular mapping of surfaces on the molecules. Recently, newer vibrational spectroscopy-based Raman mapping of cells is being used to detect cancer. Recently, Bhori and co-workers (Fig. 4.15) demonstrated use of Raman mapping in understanding cellular composition of cells—normal and cancer cells using

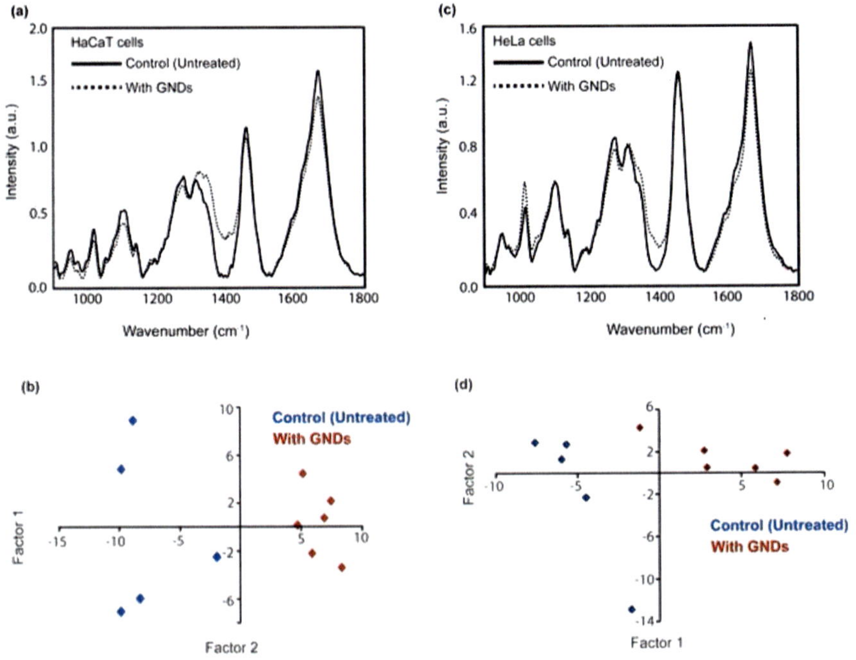

Fig. 4.15 Raman mapping of cells. The figure shows characterization of cellular composition in normal cells (HaCaT) and cancer cells (HeLa) with or without GNDs. Adapted from Bhori et al. (2020)

graphene nanodots (GNDs) treatment (Bhori et al. 2020). Such techniques provide additional information such as how cellular composition is affected in cancer cells and how treatment strategies impact such composition in real time.

References

Ajayan, P. M., et al. (2000). Single-walled carbon nanotube–polymer composites: Strength and weakness. *Advanced Materials*, *12*(10), 750–753. https://doi.org/10.1002/(SICI)1521-4095(200005)12:10%3c750::AID-ADMA750%3e3.0.CO;2-6. (John Wiley & Sons, Ltd).

Ajayan, P. M., & Zhou, O. Z. (2001). Applications of carbon nanotubes. In M. S. Dresselhaus, G. Dresselhaus, & P. Avouris (Eds.) *Carbon nanotubes: synthesis, structure, properties, and applications* (pp. 391–425). Berlin, Heidelberg: Springer. https://doi.org/10.1007/3-540-39947-x_14.

Bartelmess, J., Quinn, S. J., & Giordani, S. (2015). Carbon nanomaterials: Multi-functional agents for biomedical fluorescence and Raman imaging. *Chemical Society Reviews*. https://doi.org/10.1039/C4CS00306C. (Royal Society of Chemistry).

Bhori, M., et al. (2020). Potential of graphene nano-dots in cellular imaging and Raman mapping. *Nano*. https://doi.org/10.1142/s1793292020500988. (World Scientific Publishing Co.).

Bolskar, R. D., et al. (2003). First soluble M@C60 derivatives provide enhanced access to metallofullerenes and permit in vivo evaluation of Gd@C60[C(COOH)2]10 as a MRI contrast agent. *Journal of the American Chemical Society, 125*(18), 5471–5478. https://doi.org/10.1021/ja0340984. (American Chemical Society).

Brooks, R. A., Moiny, F., & Gillis, P. (2001). On T2-shortening by weakly magnetized particles: The chemical exchange model. *Magnetic Resonance in Medicine, 45*(6), 1014–1020. https://doi.org/10.1002/mrm.1135.

Caravan, P., et al. (1999). Gadolinium(III) chelates as MRI contrast agents: Structure, dynamics, and applications. *Chemical Reviews, 99*(9), 2293–2352. https://doi.org/10.1021/cr980440x. (American Chemical Society).

Chang, I. P., Hwang, K. C., & Chiang, C.-S. (2008). Preparation of fluorescent magnetic nanodiamonds and cellular imaging. *Journal of the American Chemical Society, 130*(46), 15476–15481. https://doi.org/10.1021/ja804253y. (American Chemical Society).

Chatterjee, N., Eom, H.-J., & Choi, J. (2014). A systems toxicology approach to the surface functionality control of graphene–cell interactions. *Biomaterials, 35*(4), 1109–1127. https://doi.org/10.1016/J.BIOMATERIALS.2013.09.108. (Elsevier).

Chaudhary, R. P., et al. (2017). Fe core-carbon shell nanoparticles as advanced MRI contrast enhancer. *Journal of functional biomaterials, 8*(4), 46. https://doi.org/10.3390/jfb8040046. (MDPI).

Chen, R. J., et al. (2003). Noncovalent functionalization of carbon nanotubes for highly specific electronic biosensors. *Proceedings of the National Academy of Sciences, 100*(9), 4984–4989. https://doi.org/10.1073/pnas.0837064100.

Chen, H., et al. (2010). Preparation and control of the formation of single core and clustered nanoparticles for biomedical applications using a versatile amphiphilic diblock copolymer. *Nano Research, 3*(12), 852–862. https://doi.org/10.1007/s12274-010-0056-y.

Chen, W., et al. (2011). Composites of Aminodextran-coated Fe3O4 nanoparticles and graphene oxide for cellular magnetic resonance imaging. *ACS Applied Materials & Interfaces, 3*(10), 4085–4091. https://doi.org/10.1021/am2009647. (American Chemical Society).

Chen, D., et al. (2017). In vivo targeting and positron emission tomography imaging of tumor with intrinsically radioactive metal-organic frameworks nanomaterials. *ACS Nano, 11*(4), 4315–4327. https://doi.org/10.1021/acsnano.7b01530. (American Chemical Society).

Cioffi, C. T., et al. (2009). A Carbon nano-onion-ferrocene donor-acceptor system: Synthesis, characterization and properties. *Chemistry—A European Journal, 15*, 4419–4427. https://doi.org/10.1002/chem.200801818.

Cong, H.-P., et al. (2010). Water-soluble magnetic-functionalized reduced graphene oxide sheets: In situ synthesis and magnetic resonance imaging applications. *Small, 6*(2), 169–173. https://doi.org/10.1002/smll.200901360. (John Wiley & Sons Ltd.).

Delhaye, M., & Dhamelincourt, P. (1975). Raman microprobe and microscope with laser excitation. *Journal of Raman Spectroscopy, 3*(1), 33–43. https://doi.org/10.1002/jrs.1250030105. (John Wiley & Sons Ltd.).

Du, F., et al. (2014). Economical and green synthesis of bagasse-derived fluorescent carbon dots for biomedical applications. *Nanotechnology, 25*, 315702. https://doi.org/10.1088/0957-4484/25/31/315702. (IOP Publishing).

Eatemadi, A., et al. (2014). Carbon nanotubes: Properties, synthesis, purification, and medical applications. *Nanoscale Research Letters, 9*(1), 393. https://doi.org/10.1186/1556-276X-9-393 .

Ebbesen, T. W., & Ajayan, P. M. (1992). Large-scale synthesis of carbon nanotubes. *Nature, 358*(6383), 220–222. https://doi.org/10.1038/358220a0.

Ember, K. J. I., et al. (2017). Raman spectroscopy and regenerative medicine: A review. *NPJ Regenerative Medicine, 2*(1), 12. https://doi.org/10.1038/s41536-017-0014-3.

Ferrari, A. C., et al. (2006). Raman spectrum of graphene and graphene layers. *Physical Review Letters, 97*(18), 187401. https://doi.org/10.1103/PhysRevLett.97.187401. (American Physical Society).

Fred, H. L. (2004). Drawbacks and limitations of computed tomography: Views from a medical educator. *Texas Heart Institute Journal, 31*(4), 345–348. Available at: https://pubmed.ncbi.nlm.nih.gov/15745283.

Frisoli, J. K., & Ph, D. (no date). High-dose gadolinium.

Frush, D., & Ogden, K. (2008). *Update course: Diagnostic radiology physics—From invisible to visible: The science and practice of X-ray imaging and radiation dose optimization.*

Goldman, L. W. (2007). Principles of CT and CT technology. *Journal of Nuclear Medicine Technology, 35*(3), 115–128. https://doi.org/10.2967/jnmt.107.042978.

Haka, A. S., et al. (2005). Diagnosing breast cancer by using Raman spectroscopy. *Proceedings of the National Academy of Sciences of the United States of America, 102*(35), 12371–12376. https://doi.org/10.1073/pnas.0501390102.

Haka, A. S., et al. (2006). In vivo margin assessment during partial mastectomy breast surgery using raman spectroscopy. *Cancer Research, 66*(6), 3317–3322. https://doi.org/10.1158/0008-5472.can-05-2815.

He, H., et al. (2013). Carbon NANOTUBES: Applications in Pharmacy and Medicine. In T. Noda (Ed.) *BioMed Research International*, Hindawi Publishing Corporation, p. 578290. https://doi.org/10.1155/2013/578290.

Hou, J., et al. (2013). A novel one-pot route for large-scale preparation of highly photoluminescent carbon quantum dots powders. *Nanoscale, 5,* 9558–9561. https://doi.org/10.1039/c3nr03444e.

Huda, W. (2014a). Kerma-Area Product in Diagnostic Radiology. *American Roentgen Ray Society, 203*(6), W565–W569. https://doi.org/10.2214/AJR.14.12513. (American Journal of Roentgenology).

Huda, W. (2014b). Understanding (and explaining) imaging performance metrics. *American Journal of Roentgenology, 203*(1), W1–W2. https://doi.org/10.2214/AJR.13.10827. (American Roentgen Ray Society).

Huda, W., & Abrahams, R. B. (2015). Radiographic techniques, contrast, and noise in X-ray imaging, pp. 126–131 (February). https://doi.org/10.2214/ajr.14.13116.

Iijima, S. (1991). Helical microtubules of graphitic carbon. *Nature, 354*(6348), 56–58. https://doi.org/10.1038/354056a0.

Jasonoff, A. (2006). Special feature—New technology for bioimaging with MRI. *Asia-Pacific Biotech News, 10*(06), 298–300. https://doi.org/10.1142/S021903030600036X. (World Scientific Publishing Co.).

Kato, H., et al. (2003). Lanthanoid endohedral metallofullerenols for MRI contrast agents. *Journal of the American Chemical Society, 125*(14), 4391–4397. https://doi.org/10.1021/ja027555+. (American Chemical Society).

Kerr, L. T., et al. (2016). Methodologies for bladder cancer detection with Raman based urine cytology. *Analytical Methods, 8*(25), 4991–5000. https://doi.org/10.1039/C5AY03300D. (The Royal Society of Chemistry).

Kim, D., et al. (2018). Recent development of inorganic nanoparticles for biomedical imaging. *ACS Central Science. American Chemical Society, 4*(3), 324–336. https://doi.org/10.1021/acscentsci.7b00574.

Krueger, A. (2008). New carbon materials: Biological applications of functionalized nanodiamond materials. *Chemistry—A European Journal, 14,* 1382–1390. https://doi.org/10.1002/chem.200700987.

Kumawat, M. K., Thakur, M., et al. (2017a). 'Graphene quantum dots for cell proliferation, nucleus imaging, and photoluminescent sensing applications. *Scientific Reports, 7*(1), 15858. https://doi.org/10.1038/s41598-017-16025-w. (Springer, US).

Kumawat, M. K., Srivastava, R., et al. (2017b). Graphene quantum dots from mangifera indica: Application in near-infrared bioimaging and intracellular nanothermometry. *ACS Sustainable Chemistry and Engineering, 5*(2), 1382–1391. https://doi.org/10.1021/acssuschemeng.6b01893.

Lalwani, G., et al. (2014). Synthesis, characterization, in vitro phantom imaging, and cytotoxicity of a novel graphene-based multimodal magnetic resonance imaging—X-ray computed tomography contrast agent. *Journal of materials chemistry. B, 2*(22), 3519–3530. https://doi.org/10.1039/C4TB00326H.

Lee, N., Choi, S. H., & Hyeon, T. (2013). Nano-sized CT contrast agents. *Advanced Materials, 25* (19), 2641–2660. https://doi.org/10.1002/adma.201300081.

Lee, T., et al. (2017). Spatially resolved Raman spectroscopy of defects, strains, and strain fluctuations in domain structures of monolayer graphene. *Scientific Reports, 7*(1), 16681. https://doi.org/10.1038/s41598-017-16969-z.

Li, H., et al. (2012). Carbon nanodots: Synthesis, properties and applications. *Journal of Materials Chemistry, 22*(46). https://doi.org/10.1039/c2jm34690g.

Li, S., et al. (2015). Characterization and noninvasive diagnosis of bladder cancer with serum surface enhanced Raman spectroscopy and genetic algorithms. *Scientific Reports, 5*(1), 9582. https://doi.org/10.1038/srep09582.

Li, Q., Chen, B., & Xing, B. (2017). Aggregation kinetics and self-assembly mechanisms of graphene quantum dots in aqueous solutions: Cooperative effects of pH and electrolytes. *Environmental Science & Technology, 51*(3), 1364–1376. https://doi.org/10.1021/acs.est.6b04178. (American Chemical Society).

Lien, Z.-Y., et al. (2012). Cancer cell labeling and tracking using fluorescent and magnetic nanodiamond. *Biomaterials, 33*(26), 6172–6185. https://doi.org/10.1016/j.biomaterials.2012.05.009.

Luo, P. G., et al. (2013). Carbon "quantum" dots for optical bioimaging. *Journal of Materials Chemistry B, 1*(16), 2116–2127. https://doi.org/10.1039/c3tb00018d.

Ma, X., et al. (2012). A functionalized graphene oxide-iron oxide nanocomposite for magnetically targeted drug delivery, photothermal therapy, and magnetic resonance imaging. *Nano Research, 5*(3), 199–212. https://doi.org/10.1007/s12274-012-0200-y.

Mahasin Alam, S.K., et al. (2014). Revealing the tunable photoluminescence properties of graphene quantum dots. *Journal of Materials Chemistry C, 2*(34), 6954–6960. https://doi.org/10.1039/c4tc01191k. (The Royal Society of Chemistry).

Manus, L. M., et al. (2010). Gd(III)-nanodiamond conjugates for MRI contrast enhancement. *Nano Letters, 10*(2), 484–489. https://doi.org/10.1021/nl903264h. (American Chemical Society).

McDonough, J., & Gogotsi, Y. (2013). Carbon onions: Synthesis and electrochemical applications. *Electrochemical Society Interface, 22,* 61–66.

McGregor, H. C., et al. (2017). Real-time endoscopic Raman spectroscopy for in vivo early lung cancer detection. *Journal of Biophotonics, 10*(1), 98–110. https://doi.org/10.1002/jbio.201500204. (John Wiley & Sons Ltd).

Mornet, S., et al. (2004). Magnetic nanoparticle design for medical diagnosis and therapy. *Journal of Materials Chemistry, 14*(14), 2161–2175. https://doi.org/10.1039/B402025A. (The Royal Society of Chemistry).

Mu, Q., et al. (2012). Size-dependent cell uptake of protein-coated graphene oxide nanosheets. *ACS Applied Materials and Interfaces, 4*(4), 2259–2266. https://doi.org/10.1021/am300253c. (American Chemical Society).

Oza, G., Oza, K., Pandey, S., Shinde, S., Mewada, A., Thakur, M., Sharon, M., & Sharon, M. (2015). A green route towards highly photoluminescent and cytocompatible carbon dot synthesis and its separation using sucrose density gradient centrifugation. *Journal of Fluorescence, 25*(1), 9–14. https://doi.org/10.1007/s10895-014-1477-x.

Padalkar, M. V., & Pleshko, N. (2015). Wavelength-dependent penetration depth of near infrared radiation into cartilage. *The Analyst, 140*(7), 2093–2100. https://doi.org/10.1039/c4an01987c.

Pan, B. D., et al. (2010). Hydrothermal route for cutting graphene sheets into blue-luminescent graphene quantum dots (pp. 734–738). https://doi.org/10.1002/adma.200902825.

Pan, J., et al. (2018). Gd3+ -doped MoSe2 nanosheets used as a theranostic agent for bimodal imaging and highly efficient photothermal cancer therapy. *Biomaterials Science, 6*(2), 372–387. https://doi.org/10.1039/C7BM00894E. (The Royal Society of Chemistry).

Pandey, S., et al. (2014). Synthesis of mesoporous silica oxide/C-dot complex (meso-SiO 2/ C-dots) using pyrolysed rice husk and its application in bioimaging. *RSC Advances*, *4*(3). https://doi.org/10.1039/c3ra45227a.

Pandey, S., Thakur, M., et al. (2013). Carbon dots functionalized gold nanorod mediated delivery of doxorubicin: Tri-functional nano-worms for drug delivery, photothermal therapy and bioimaging. *Journal of Materials Chemistry B*. https://doi.org/10.1039/c3tb20761g. (The Royal Society of Chemistry).

Pandey, S., Mewada, A., et al. (2013). Cysteamine hydrochloride protected carbon dots as a vehicle for the efficient release of the anti-schizophrenic drug haloperidol. *RSC Advances, 3* (48), 26290. https://doi.org/10.1039/c3ra42139b.

Puvvada, N., et al. (2012). Synthesis of biocompatible multicolor luminescent carbon dots for bioimaging applications. *Science and Technology of Advanced Materials, 13*, 045008. https://doi.org/10.1088/1468-6996/13/4/045008.

Sandborg, M., et al. (1995). The physical performance of different X-ray contrast agents: Calculations using a Monte Carlo model of the imaging chain. *Physics in Medicine and Biology, 40*(7), 1209–1224. https://doi.org/10.1088/0031-9155/40/7/005. (IOP Publishing).

Servant, A., et al. (2016). Gadolinium-functionalised multi-walled carbon nanotubes as a T1 contrast agent for MRI cell labelling and tracking. *Carbon, 97*, 126–133. https://doi.org/10.1016/j.carbon.2015.08.051.

Shi, X., et al. (2013). Graphene-based magnetic plasmonic nanocomposite for dual bioimaging and photothermal therapy. *Biomaterials, 34*(20), 4786–4793. https://doi.org/10.1016/j.biomaterials.2013.03.023.

Shi, J., et al. (2014). A tumor-targeting near-infrared laser-triggered drug delivery system based on GO@Ag nanoparticles for chemo-photothermal therapy and X-ray imaging. *Biomaterials, 35* (22), 5847–5861. https://doi.org/10.1016/j.biomaterials.2014.03.042.

Shokrollahi, H. (2013). Contrast agents for MRI. *Materials Science & Engineering C, 33*(8), 4485–4497. https://doi.org/10.1016/j.msec.2013.07.012. (Elsevier B.V.).

Sinharay, S., & Pagel, M. D. (2016). Advances in magnetic resonance imaging contrast agents for biomarker detection. *Annual Review of Analytical Chemistry. Annual Reviews, 9*(1), 95–115. https://doi.org/10.1146/annurev-anchem-071015-041514.

Sitharaman, B., & Wilson, L. J. (2006). Gadonanotubes as new high-performance MRI contrast agents. *International journal of nanomedicine*. *1*(3), 291–295. Available at: https://pubmed.ncbi.nlm.nih.gov/17717970. (Dove Medical Press).

Sonkar, S. K., et al. (2012). Water soluble carbon nano-onions from wood wool as growth promoters for gram plants. *Nanoscale*, 7670–7675. https://doi.org/10.1039/c2nr32408c.

Sun, H., et al. (2013a). The effects of composition and surface chemistry on the toxicity of quantum dots. *Journal of Materials Chemistry B, 1*, 6485–6494. https://doi.org/10.1039/C3TB21151G.

Sun, Y., et al. (2013b). Highly efficient enrichment of radionuclides on graphene oxide-supported polyaniline. *Environmental Science & Technology, 47*(17), 9904–9910. https://doi.org/10.1021/es401174n. (American Chemical Society).

Sun, B., Cui, R., Huang, H., & Guo, X. (2015). Novel Carbon Nanohybrids As Highly Efficient Magnetic Resonance Imaging Contrast Agents. (2015). *ECS Meeting Abstracts*. The Electrochemical Society. https://doi.org/10.1149/ma2015-01/7/816.

Sun, B., et al. (2017). In situ synthesis of graphene oxide/gold nanorods theranostic hybrids for efficient tumor computed tomography imaging and photothermal therapy. *Nano Research, 10* (1), 37–48. https://doi.org/10.1007/s12274-016-1264-x.

Talapatra, S., et al. (2005). Irradiation-induced magnetism in carbon nanostructures. *Physical Review Letters, 95*(9), 97201. https://doi.org/10.1103/PhysRevLett.95.097201. (American Physical Society).

Tao, H., et al. (2012). In vivo NIR fluorescence imaging, biodistribution, and toxicology of photoluminescent carbon dots produced from carbon nanotubes and graphite. *Small, 8*(2), 281–290. https://doi.org/10.1002/smll.201101706. (Weinheim an der Bergstrasse, Germany).

Thakur, M., et al. (2014). Antibiotic conjugated fluorescent carbon dots as a theranostic agent for controlled drug release, bioimaging, and enhanced antimicrobial activity. *Journal of drug delivery*, 282193. https://doi.org/10.1155/2014/282193.

Thakur, M., et al. (2016). Milk-derived multi-fluorescent graphene quantum dot-based cancer theranostic system. *Materials Science and Engineering C*, 67. https://doi.org/10.1016/j.msec.2016.05.007.

Thomsen, H. S. (2007). ESUR guideline: gadolinium-based contrast media and nephrogenic systemic fibrosis. *European Radiology*, *17*(10), 2692–2696. Available at: http://www.ncbi.nlm.nih.gov/entrez/query.fcgi?cmd=Retrieve&db=PubMed&dopt=Citation&list_uids=17977076.

Tian, P., et al. (2018). Graphene quantum dots from chemistry to applications. *Materials Today Chemistry, 10,* 221–258. https://doi.org/10.1016/J.MTCHEM.2018.09.007. Elsevier.

Tripathi, S. K., et al. (2013). Functionalized graphene oxide mediated nucleic acid delivery. *Carbon, 51*(1), 224–235. https://doi.org/10.1016/j.carbon.2012.08.047. Elsevier Ltd.

Wang, Y. ying, et al. (2008). Raman studies of monolayer graphene: The substrate effect. *The Journal of Physical Chemistry C, 112*(29), 10637–10640. https://doi.org/10.1021/jp8008404. (American Chemical Society).

Wang, X., et al. (2010). Bandgap-like strong fluorescence in functionalized carbon nanoparticles. *Angewandte Chemie—International Edition, 49,* 5310–5314. https://doi.org/10.1002/anie.201000982.

Xie, W., et al. (2018). Shape-, size- and structure-controlled synthesis and biocompatibility of iron oxide nanoparticles for magnetic theranostics. *Theranostics, 8*(12), 3284–3307. https://doi.org/10.7150/thno.25220. (Ivyspring International Publisher).

Xu, X. et al. (2004). Electrophoretic analysis and purification of fluorescent single-walled carbon nanotube fragments (pp. 12736–12737).

Yan, Y., et al. (2019). Recent advances on graphene quantum dots: From chemistry and physics to applications. *Advanced Materials, 31*(21), 1808283. https://doi.org/10.1002/adma.201808283. (John Wiley & Sons Ltd).

Yang, K., et al. (2012). Multimodal imaging guided photothermal therapy using functionalized graphene nanosheets anchored with magnetic nanoparticles. *Advanced Materials, 24*(14), 1868–1872. https://doi.org/10.1002/adma.201104964. (John Wiley & Sons Ltd).

Ye, R., et al. (2013). Coal as an abundant source of graphene quantum dots. Nature communications, *4*, 2943. https://doi.org/10.1038/ncomms3943.

Yu, S.-B., & Watson, A. D. (1999). Metal-based x-ray contrast media. *Chemical Reviews, 99*(9), 2353–2378. https://doi.org/10.1021/cr980441p. (American Chemical Society).

Zhang, M., et al. (2013). Graphene oxide based theranostic platform for T1-weighted magnetic resonance imaging and drug delivery. *ACS Applied Materials & Interfaces, 5*(24), 13325–13332. https://doi.org/10.1021/am404292e. (American Chemical Society).

Zhang, H., et al. (2015). Graphene oxide-BaGdF5 nanocomposites for multi-modal imaging and photothermal therapy. *Biomaterials, 42,* 66–77. https://doi.org/10.1016/j.biomaterials.2014.11.055.

Zhang, B., et al. (2016). Interactions of graphene with mammalian cells: Molecular mechanisms and biomedical insights. *Advanced Drug Delivery Reviews, 105,* 145–162. https://doi.org/10.1016/J.ADDR.2016.08.009. (Elsevier).

Zhang, Yixue, et al. (2017). Hydrophilic graphene oxide/bismuth selenide nanocomposites for CT imaging, photoacoustic imaging, and photothermal therapy. *Journal of Materials Chemistry B, 5*(9), 1846–1855. https://doi.org/10.1039/C6TB02137A. (The Royal Society of Chemistry).

Zhang, G., et al. (2018). A tailored nanosheet decorated with a metallized dendrimer for angiography and magnetic resonance imaging-guided combined chemotherapy. *Nanoscale, 10* (1), 488–498. https://doi.org/10.1039/C7NR07957E. (The Royal Society of Chemistry).

Zhao, A., et al. (2015). Recent advances in bioapplications of C-dots. *Carbon, 85,* 309–327. https://doi.org/10.1016/j.carbon.2014.12.045. (Elsevier Ltd).

Zheng, X. T., et al. (2015). Glowing graphene quantum dots and carbon dots: Properties, syntheses, and biological applications. *Small,* 1620–1636. https://doi.org/10.1002/smll.201402648.

Zhou, J., et al. (2012). Facile synthesis of fluorescent carbon dots using watermelon peel as a carbon source. *Materials Letters*, *66*(1), 222–224. https://doi.org/10.1016/j.matlet.2011.08.081 . (Elsevier B.V.).

Zhou, C., Wang, J., & Szpunar, J. A. (2014). X-ray chemical imaging and the electronic structure of a single nanoplatelet Ni/graphene composite. *Chemical Communications, 50*(18), 2282–2285. https://doi.org/10.1039/C3CC47008C. (The Royal Society of Chemistry).

Zhu, S., et al. (2015). The photoluminescence mechanism in carbon dots (graphene quantum dots, carbon nanodots, and polymer dots): current state and future perspective. *Nano Research, 8,* 355–381. https://doi.org/10.1007/s12274-014-0644-3.

Chapter 5
Graphene-Based Nanomaterials in Cancer Therapy

5.1 Introduction

Since the past two decades, graphene-based nanomaterials have been envisaged for biological applications since the discovery of graphene (Novoselov et al. 2004; Geim and Novoselov 2007). With the advent of the new members of the carbon family and its expansion (Fig. 5.1a), their interesting properties have led toward exciting new applications in medicine and biology.

In light of such graphene-based nanomaterials, apart from graphene oxide (GO), carbon nanomaterials such as the graphene quantum dots (GQDs) or carbon dots (CDs) or polymer dots (PDs) will be discussed in detail in terms of their structure, property, and application relationships in biology (Zheng et al. 2015; Chen et al. 2017; Yan et al. 2019). The properties such as molecular size, high surface-to-volume ratio, biocompatibility, and their response to external stimulus (e.g., light), photo-stability, and photo-tunability are important in applications involving therapy and diagnosis of cancer. Current approaches toward cancer therapy are limited by side effects and toxicity from chemotherapeutic drugs and radiotherapy. Graphene nanomaterials hold potential for targeted drug delivery to the cancer cells with minimal toxicity to normal cells and tissues. In addition, they also provide innovative non-invasive techniques to decrease cancer cell progression and disease (Liu et al. 2011; Hong et al. 2015; Zheng et al. 2015; Mohajeri et al. 2018).

A typically a mammalian cell is few tens of microns in size. For applications such as cargo delivery or cellular imaging, the nanomaterials need to be smaller than the cell itself. The graphene nanomaterials, in particular, vary in their size up to hundreds of nanometers. These nanomaterials are quite attractive in biological applications because their size is similar to cellular biomolecules (e.g., DNA, proteins, microtubules). Furthermore, they are taken up by the cells (Fig. 5.1b) owing to the nanomaterials' properties such as inertness, surface charge, and chemical composition making them an ideal candidates for delivery of therapeutic agents (Tian et al. 2018). Some of the important properties of nanomaterials in

R. Srivastava et al., *Next Generation Graphene Nanomaterials for Cancer Theranostic Applications*,
https://doi.org/10.1007/978-981-33-6303-8_5

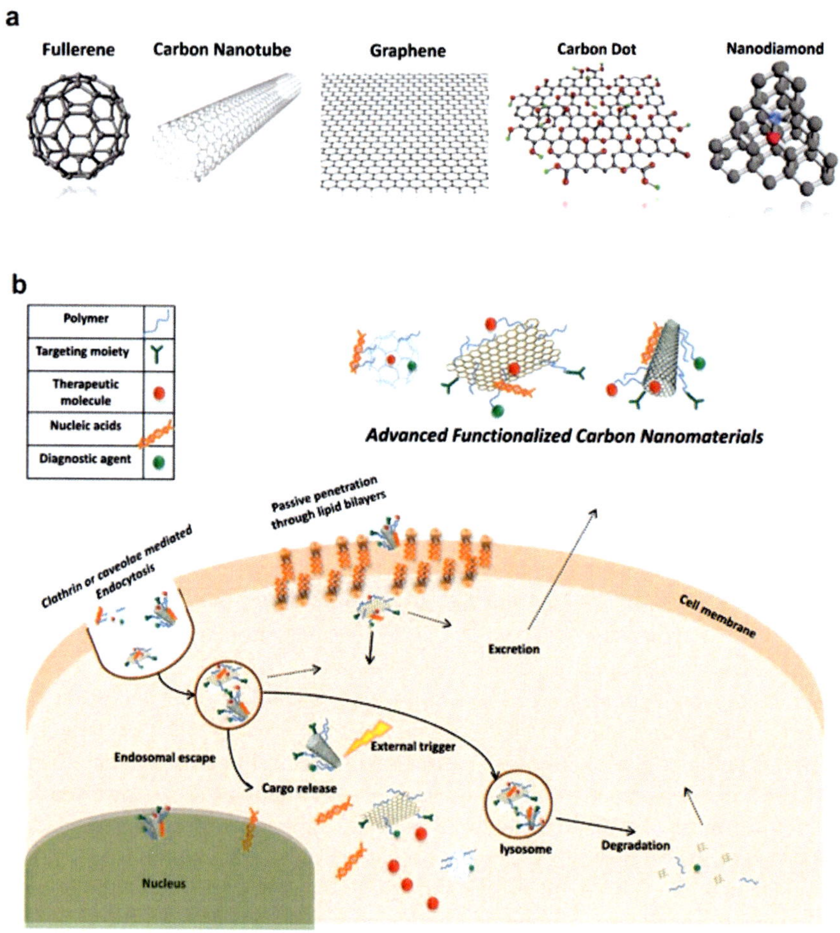

Fig. 5.1 **a** Major types of carbon nanomaterials used in cancer therapy. Reprinted (adapted) with permission from (Hong et al. 2015). Copyright (2015) American Chemical Society, **b** an illustration showing example of cargo functionalization and delivery mechanism inside cell. Adapted from (Mohajeri et al. 2018)

order to use as therapeutic delivery agents are shown in Fig. 5.2. From the cancer diagnostics perspective, an ideal imaging fluorescent probe should possess high photoluminescence quantum yield to differentiate the real signal from the background noise and at the same time the optical signal needs to be long enough to be detected. Since the graphene nanomaterials possess excellent physicochemical properties such as inherent photoluminescence and optical stability, they are also a very good candidate for imaging agents.

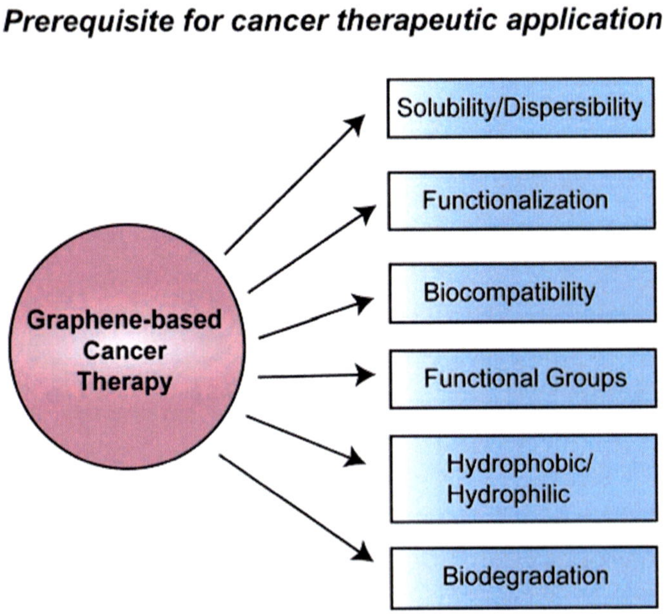

Fig. 5.2 Some prerequisite properties of carbon nanomaterials relevant to cargo delivery applications in cancer therapy

5.2 Payload Delivery

The graphene nanomaterials need to interact with the cells in order to realize its applicability (e.g., cargo delivery) to the cancer cells. Typically, these nanomaterials are in the range of sizes extending from few nanometers in their lateral dimensions to as high as few hundreds of nanometers. For example, the GQDs/CDs/PDs are zero-dimensional nanomaterials and their sizes (in non-agglomerated form) can be 1–10 nm and average thickness ~1–2 nm while the size of CNDs can range from 2 to 10 nm. Similarly, GO can extend further in their lateral dimensions from 10^2 to 10^3 nm range. These nanomaterials possess excellent cargo (drugs/genes/aptamers/proteins) loading capability owing to their large surface area, functional groups available to form chemical bonds with a stable interaction arising through π-π stacking, hydrophobic, electrostatic, or physisorption (Hong et al. 2015). The term cargo here refers to any pharmacologically active molecules (e.g., chemotherapeutic drugs, antibiotics) or biopolymers such as DNA/RNA or even proteins or polypeptides. The main reason behind cargo delivery to the cancer cells is to specifically target these cells with anticancer drugs, perform localized delivery at the site for effective therapy, reduce the amount of drug dosage and abrogate drug-related side effects. Another aspect of cancer is known as the enhanced permeability and retention (EPR) effect (Fig. 5.3). The blood vessels carrying nutrients

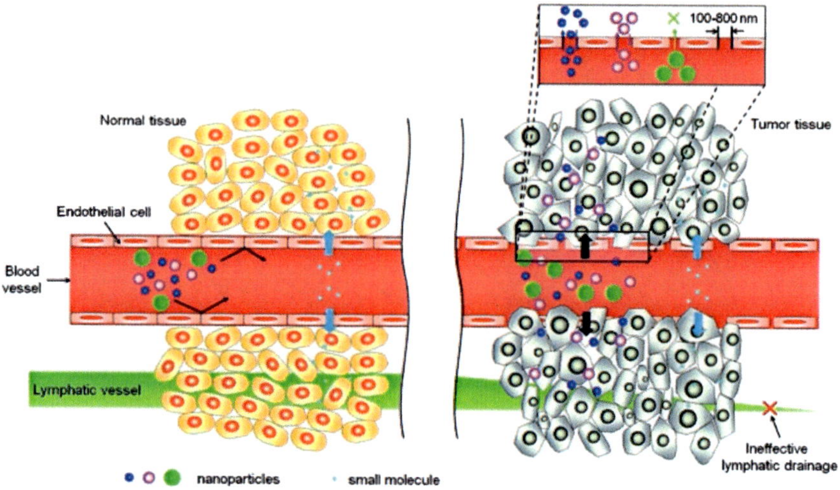

Fig. 5.3 Schematic showing comparison of nanoparticles (different sizes) diffusing and subsequently leaking in corresponding cell matrices or tissue. The cancerous tissue (on the right) shows enhanced permeation and retention (EPR) effect with accumulation of nanoparticles in contrast to healthy tissue. Adapted from (Sun et al. 2014)

and oxygen to the tumorous tissue are abnormal and leaky. This causes drainage and high accumulation of the nanomaterials near the tumorous tissues and referred to as EPR effect.

5.2.1 *High-Molecular Weight Cargoes*

The high-molecular cargoes comprise of macromolecules such as peptides, proteins, aptamers, DNA/RNA, short interfering RNA (siRNA), and plasmid DNA. For cancer therapies involving transportation of biomolecules (e.g., DNA/RNA) into the cells is performed with an aim to change the disease's genetic pathway and hamper the cellular proliferation of the abnormal cell. The major problem associated with transportation of these biomacromolecules is that they are susceptible to degradation in the body either through nuclease-activity or proteolysis before reaching their target. In such a case, the effective lifetime of the biomacromolecules, such as DNA/RNA/peptides, is significantly reduced and affects the delivery process. To address these challenges, graphene-based delivery system has been researched for delivering the cargo (biomacromolecules) into the cancer cell. In principle, the nanomaterial needs to form stable complex with biomacromolecule, it should preserve the biomacromolecule from lysis enzymes, and efficiently promotes the passage across the membranes. Let us see some examples where such nanomaterials have been used toward gene delivery in cancer cells and their outcomes so far (Table 5.1).

Table 5.1 Additional information of graphene-based nanomaterials in gene and protein therapy shown in the case studies

Carrier type	Size range	Charge	Cargo	Cell lines/ animal models	Remarks	References
siRNA-C-dots-PEI	~5 nm	~12 mV	siRNA	MGC-803	Downregulation of surviving protein followed by apoptosis and G1 phase cell cycle arrest in gastric cancer cells	Q. Wang et al. (2014)
Folic acid-C-dots-PEI-siRNA	~175 nm	−9 mV	siRNA	H460 cells	High intracellular accumulation of the complex and gene silencing	Wu et al. (2016)
P-dots (PDs)	33 nm	−23 mV	siRNA	MCF-7	High complexing capability with siRNA and successful suppression of EG5 gene	Wang et al. (2019)
CDs-EGFP	200–250 nm complex	−35 to −43 mV	EGFP protein	HeLa	Intracellular protein delivery in cells	Zhang et al. (2016)

For lung cancer therapy, especially non-small cell lung cancers (NSCLC), drug resistance is a major problem after longer chemotherapy sessions. Therefore, gene silencing is one of the proposed ways to reduce drug resistance by silencing the expression of the oncogene inside cells. In this respect, small interference RNA (siRNA) have been proposed for gene silencing precisely due to ease of design of the sequence, safety in terms of gene modification, and their effectiveness in reducing the heterogenous cancer cell progression. However, their delivery into cells is seriously hindered by nuclease degradation inside cells. Wang and co-workers synthesized a water-soluble and biocompatible C-dots (CDs) complex along with PEI and delivered siRNA into human gastric cells (Fig. 5.4a). What they observed was that the intracellular concentration was increased significantly thereby silencing the survivin gene (~6%) (Q. Wang et al. 2014). Their study shows that a simple microwave synthesized CDs can be surface modified by long biomolecular structures (siRNA) and delivered safely into the cells. The CDs in essence provided structural integrity and resistance to enzymatic access and thereby reducing degradation of the RNA. Such delivery system can advance the field of non-viral vector-based delivery systems for siRNA therapeutics in cancer. In separate study, a folic acid conjugated CDs system was developed for targeted therapy (Wu et al.

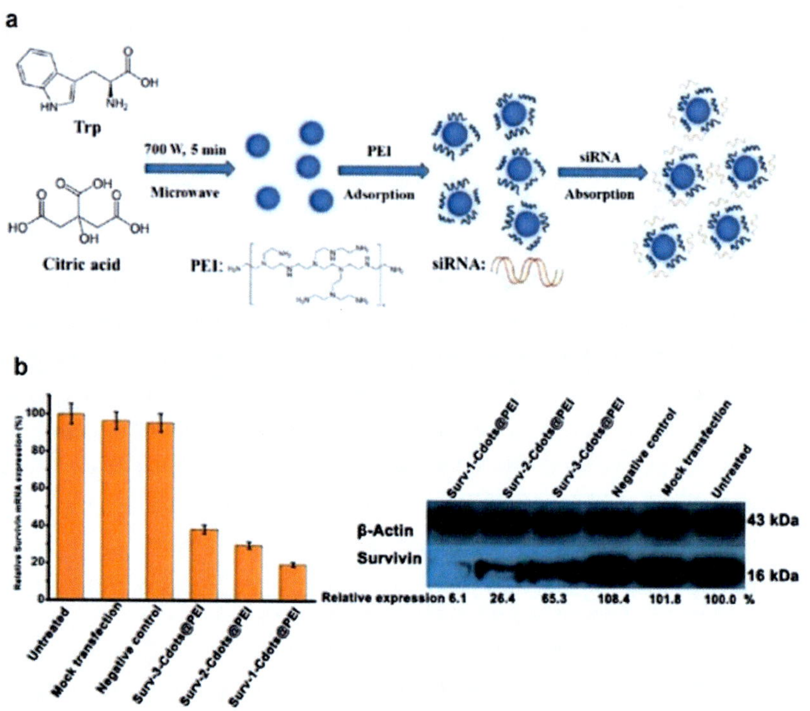

Fig. 5.4 **a** Showing synthesis of siRNA-C-dots-PEI cargo, **b** Efficiency of gene silencing for the same cargo against mRNA and protein of survivin gene. Adapted from (Q. Wang et al. 2014)

2016). They developed multifunctional CDs which were passivated by reduced PEI and attached with folic acid (fc) on the surface (Fig. 5.5a). The complex was then used to deliver multiple siRNA (e.g., EGFR and cyclin B1) inside cells. They observed that the siRNA selectively accumulated inside lung cancer cells and caused effective gene silencing. With this approach, one can see that the CDs are capable of delivering more than one type of siRNA molecules while maintaining their functionality in cancer gene silencing (Fig. 5.5b).

Moving on to some other types of carbon nanomaterials that were also evaluated for their efficiency in gene delivery. For instance, polymer dots (P-dots or PDs) were used as a platform for conjugation of functional lipids and transferring siRNA in cells (Fig. 5.6). Using electrostatic interactions, the cationic lipid (G0C14) was attached on polymer (PFO) to later form P-dots of ~30 nm size (Fig. 5.6a, b). Such P-dots were not only biocompatible but were also capable of silencing EG5 gene in MCF-7 cell lines. The gene silencing effect using P-dots delivery system was comparable to the commercial transfection agent lipofectamine 2000 (Fig. 5.6c). Thus, a whole new possibility of using a different type of carbon nanomaterial was shown for gene delivery.

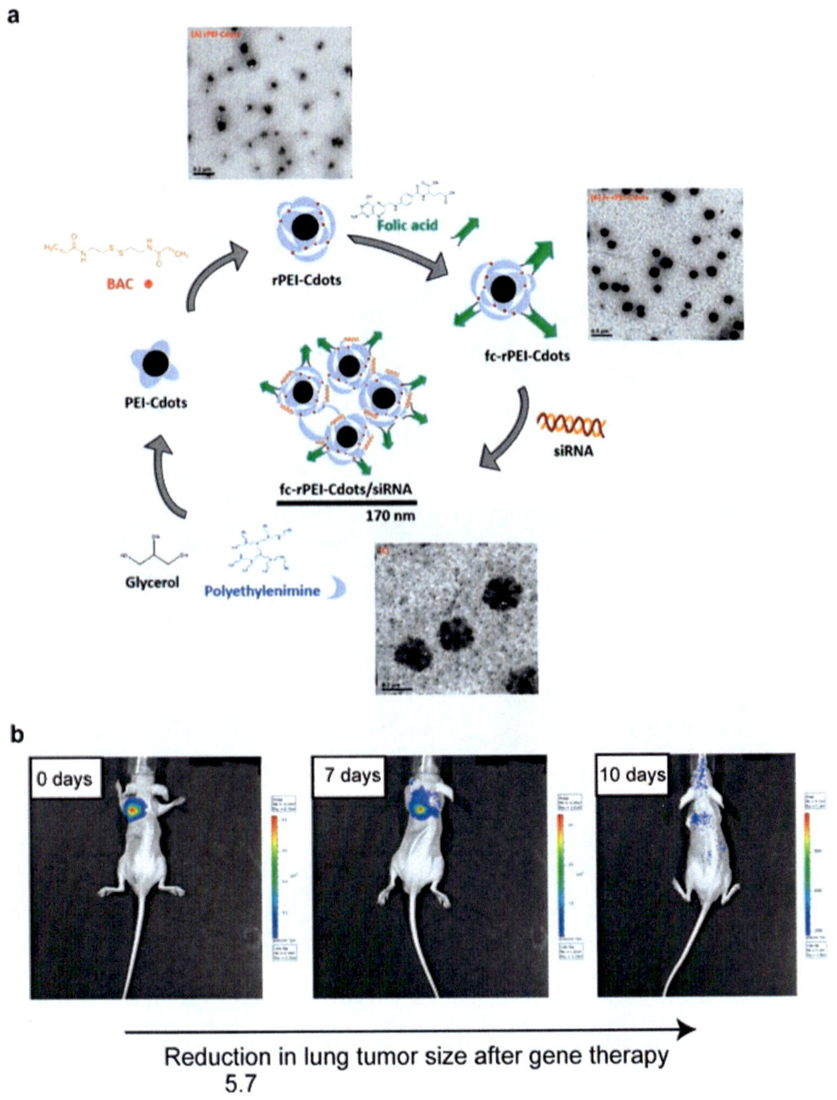

Fig. 5.5 **a** Schematic showing synthesis of theranostic agent for gene therapy. The complex, referred to as folic acid conjugated reduced-PEI-C dots is attached to siRNA to form fc-rPEI-C-dots/siRNA complex, **b** In vivo analysis of luciferase-expressing H460 lung cancer cells after treatment with form fc-rPEI-C-dots/siRNA complex. The complex was introduced through oral route and significant reduction in the tumor was observed in 10 days post-therapy. Adapted from Wu et al. (2016)

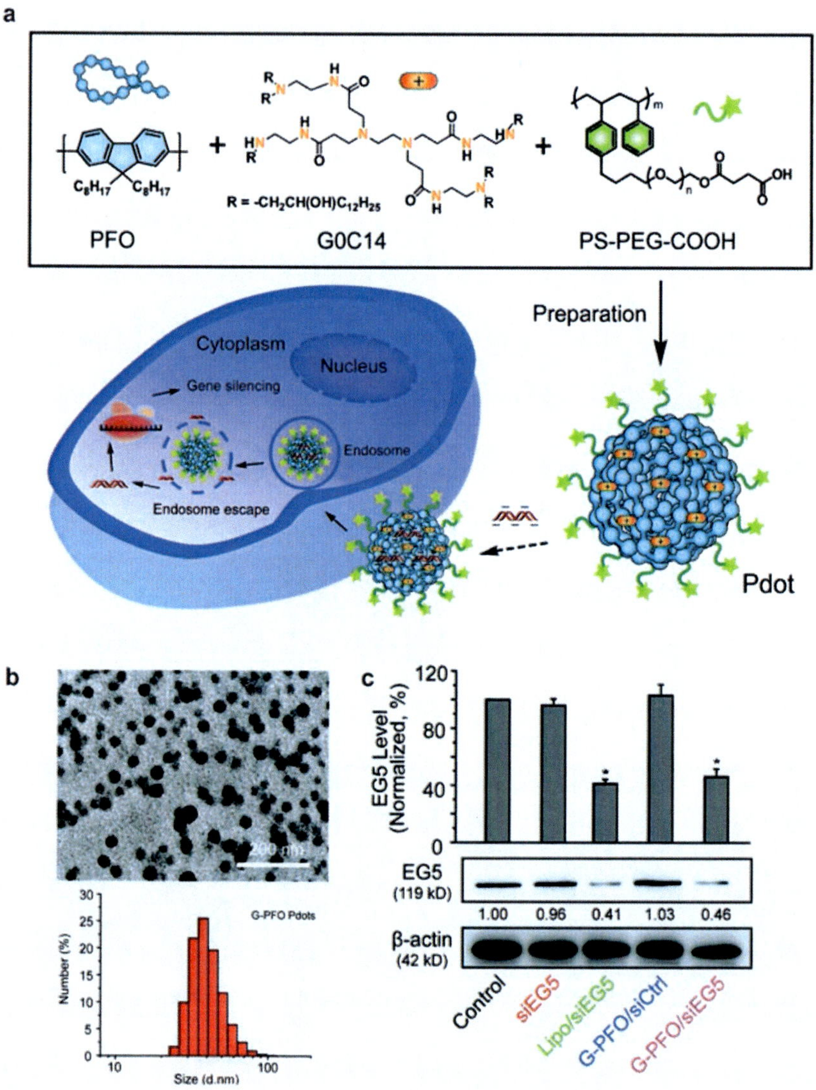

Fig. 5.6 **a** Schematic showing synthesis of cationic polymers dots (G-PFO Pdot), **b** TEM imaging and size analysis of the G-PFO Pdots showing and an average of ~30 nm, **c** Western botting results showing quantitative analysis of EG5 protein from MCF-7 cells showing effective gene silencing effect by GFO/siEG5. Reproduced from Ref. Wang et al. (2019) with permission from the Centre National de la RechercheScientifique (CNRS) and The Royal Society of Chemistry

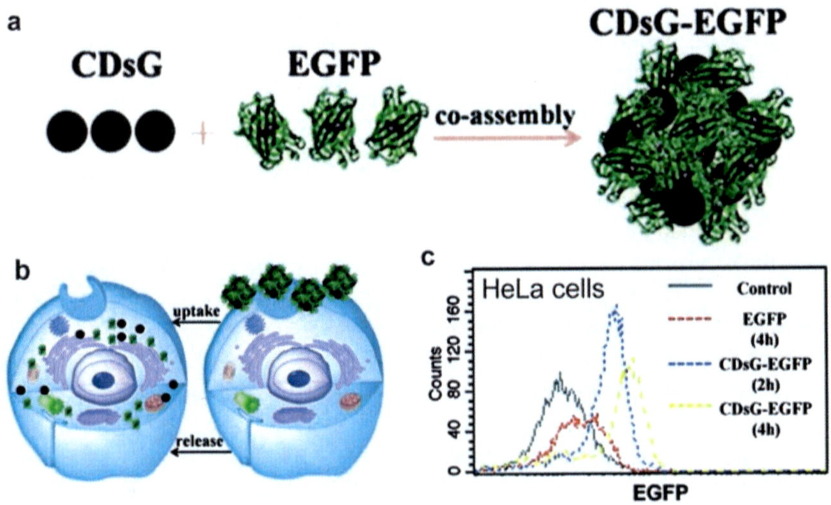

Fig. 5.7 **a** Assembly of enhanced green fluorescent protein with glutamic acid synthesized CDs, **b** Schematic showing cellular uptake of CDsG-EGFP complex, **c** Quantitative analysis showing enhanced permeability of CDsG-EGFP complex in HeLa cells Reproduced from Ref. Zhang et al. (2016) with permission from The Royal Society of Chemistry

Furthermore, carbon dots or (CDs) were also developed for transport of protein molecules. The enhanced green fluorescent protein (EGFP) was used as a model to analyze protein transferring capability of CDs. The CDs were synthesized from glutamic acid (CDsG) and assembled with EGFP to form CDsG-EGFP complex (Fig. 5.7a). The complex internalized very efficiently in HeLa cells showing higher cellular localization of the complex (Fig. 5.7b–c). This study is particularly interesting and demonstrates that not only gene delivery (as shown before) but even protein molecules can be delivered into the cells using CDs. The CDs protected the protein from proteolysis degradation and us the protein, in principle, may be replaced with any functional proteins for intracellular delivery. The attachment method was based on a simple electrostatic interaction and did not cause detrimental effect on the structure of the protein. Thus the CDs provided new avenues for intracellular macromolecule delivery system for cellular engineering.

5.2.2 *Low-Molecular Weight Cargoes*

The low-molecular weight cargoes comprise of chemotherapeutics drugs, antibiotics, and nanoimaging probes (e.g., fluorescent dyes, semiconductor quantum dots). These nanomaterials are used for cancer therapy due to their many properties mentioned in Fig. 5.2. Among them, large surface area and surface functionalization property

(Chen et al. 2019) make them attractive for chemotherapeutic drug delivery. At the same time, as the main aim of cargo delivery of therapeutics is to maximize the accumulation of drug at the required site and decrease toxicity. At this point, we also introduce a term—'theranostics' a very recent and booming field with respect to cancer therapeutics.

In principle, a theranostic nanoparticle or its complex can perform multiple biofunctions. They are multifunctional agents with an ability to perform diagnosis as well treatment of the cancer cells. These types of nanoparticle complexes are sometimes also referred to as 'theragnostic' agents (Fig. 5.8). Most of the applications involving such small or low-molecular weight payload are with the graphene nanomaterials such as carbon nanotubes, graphene, GO, CDs, GQDs, PDs in the context of cancer (Xie et al. 2010; Liu et al. 2011; Sun et al. 2014; Hong et al. 2015). Hence, many types of graphene-based nanomaterials are explored, and we will see representative case studies of such innovative therapeutic interventions.

During initial studies toward application of GQDs for drug delivery, a simple system of attachment of anticancer drug to GQDs surface was explored by Wang and co-workers (X. Wang et al. 2014). They prepared and modified folic acid (FA) on the surface of GQDs and attached a chemotherapeutic drug doxorubicin (DOX) to form DOX-FA-GQDs complex. With such complex they could show anticancer activity of the complex in in vitro cancer cells. They observed that the therapeutic activity was low compared to the bare drug but the toxicity of GQDs was very low. However, the drug release was considerably decreased leading to sustained release profile. From research point of view, a few important points can be focused on the biocompatibility of GQDs, their role as a cargo attachment, and potential carrier for chemotherapeutic drugs. Similarly, as shown in Fig. 5.9, Liang and co-workers (2019) used GQDs along with DOX and impregnated them in a polymer coating of PLGA. Later the complex was coated with BSA to form

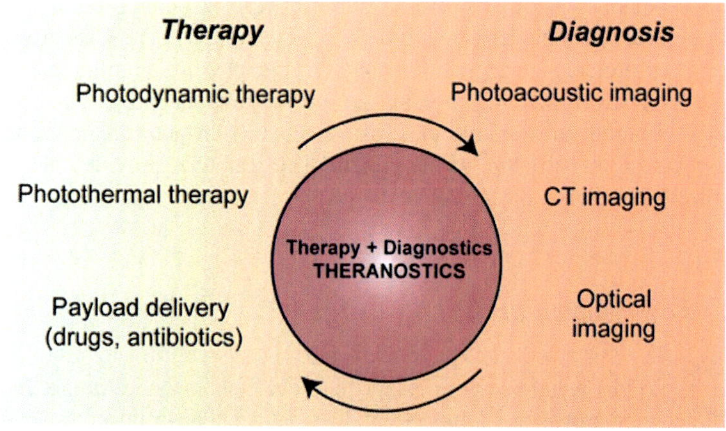

Fig. 5.8 Schematic showing most widely used examples for theranostic application in cancer therapy

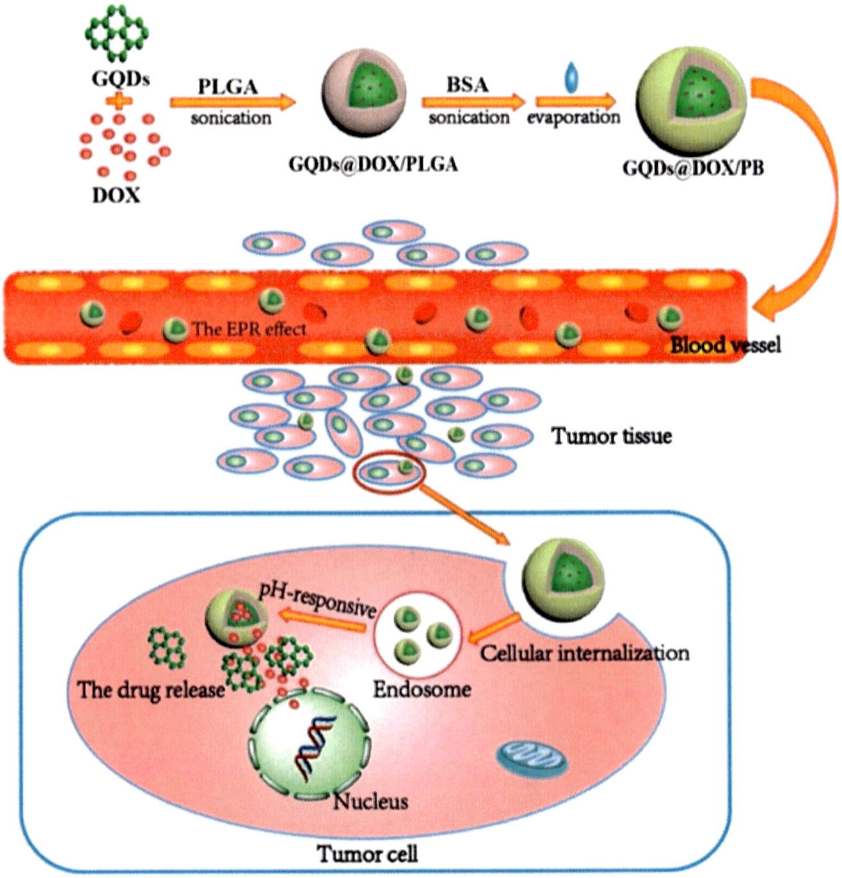

Fig. 5.9 Schematic showing synthesis of GQDs loaded DOX/PB nanoparticles. The complex is then internalized and used for cancer therapy in HeLa cells. Adapted from Liang et al. (2019)

GQDs@DOX/PB nanoparticles and used for sustained release of DOX in HeLa cells.

Later, many researchers focused on using GQDs as drug delivery platform for similar drug delivery approaches. Khodadadei et al. used methotrexate (MTX), another anticancer drug by attaching it to nitrogen-doped GQDs (N-GQDs) of ~10 nm and studied its therapeutic efficacy (Khodadadei et al. 2017). With their system, they observed higher cytotoxic effect from MTX-N-GQDs compared to bare MTX on human breast cancer cells (Fig. 5.10). Meanwhile, Wang and co-workers synthesized hollow-structured carbon dots (HCDs) and used it to deliver DOX into lung cancer cells in vitro (Wang et al. 2013). Meanwhile, other studies (Kong et al. 2018) showed that the CDs-DOX could cause cancer cell death by causing cellular apoptosis (Fig. 5.11).

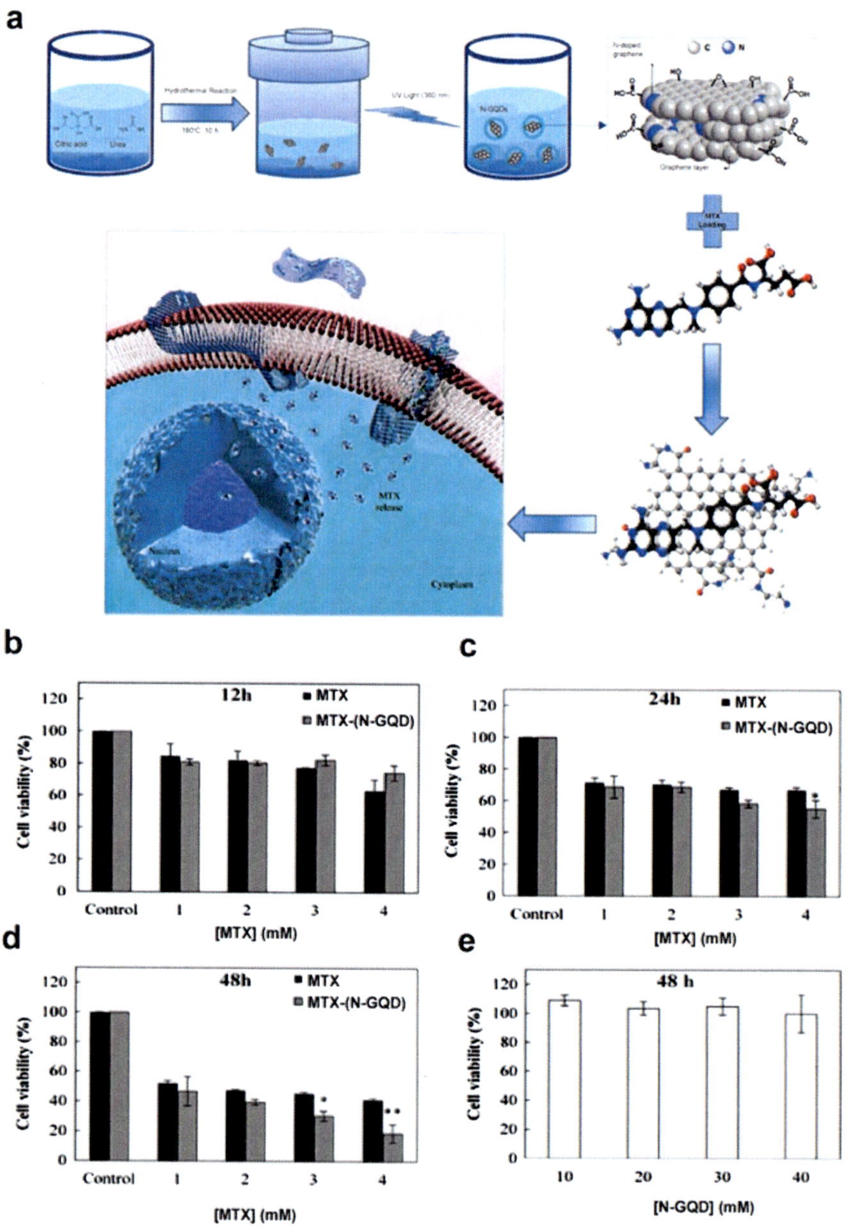

Fig. 5.10 **a** Schematic for synthesis for nitrogen-doped N-GQDs and intracellular transport of methotrexate drug (MTX), MCF-7 cell viability analysis of N-GQDs-MTX after **b** 12 h, **c** 24 h, **d** 48 h, **e,** biocompatibility of bare N-GQDs. Adapted from Khodadadei et al. (2017)

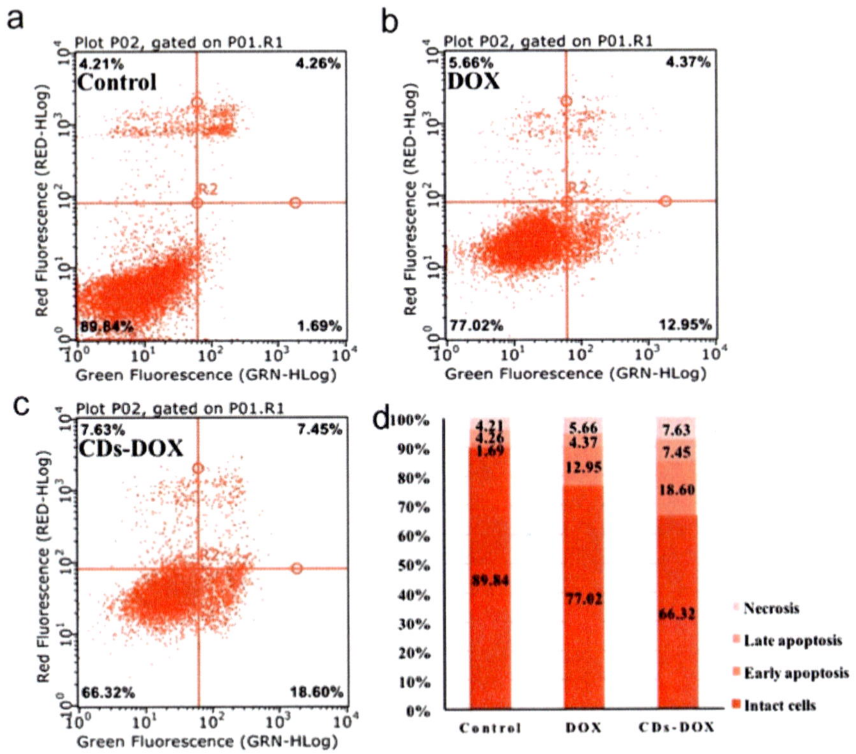

Fig. 5.11 Flow cytometry analysis of MCF-7 cells showing apoptosis as a major cell death cause in cancer cells post-treatment (24 h, drug concentration: 0.5 μg mL^{-1}). **a** Control (Untreated), **b** bare DOX, **c** CDs-DOX complex, **d** showing summary of apoptosis events in the cells. Adapted from Kong et al. (2018)

Other approaches that used CDs as drug delivery vehicle included using oxaliplatin (Oxa) as CD-Oxa complex. The complex aimed at utilizing the fluorescence property of CDs and anticancer property of Oxa (Fig. 5.12). This approach was used for simultaneous therapeutic and tracking cancer cells, sort of theranostic application in HeLa cells and mouse models with H22 liver cancer. Using this technique, one could monitor the distribution of chemotherapeutic drugs and enable specific customization for the injection and dosage control.

In another approach, an elegant way of ON/OFF switch using CDs was demonstrated by Zhou and co-workers (Zhou et al. 2013). Mesoporous silica was carbon-gated using CDs and the complex was filled with anticancer drug DOX (Fig. 5.13). The complex was biocompatible in nature. The resultant complex was pH-sensitive and could release DOX from CDs@MSPs complex in a pH-dependent manner. Effectively, with such study one can expect a pH-sensitive biovehicle for cancer therapeutics. In another approach, Feng and co-workers demonstrated a tumor microenvironment specific complex (Fig. 5.14).

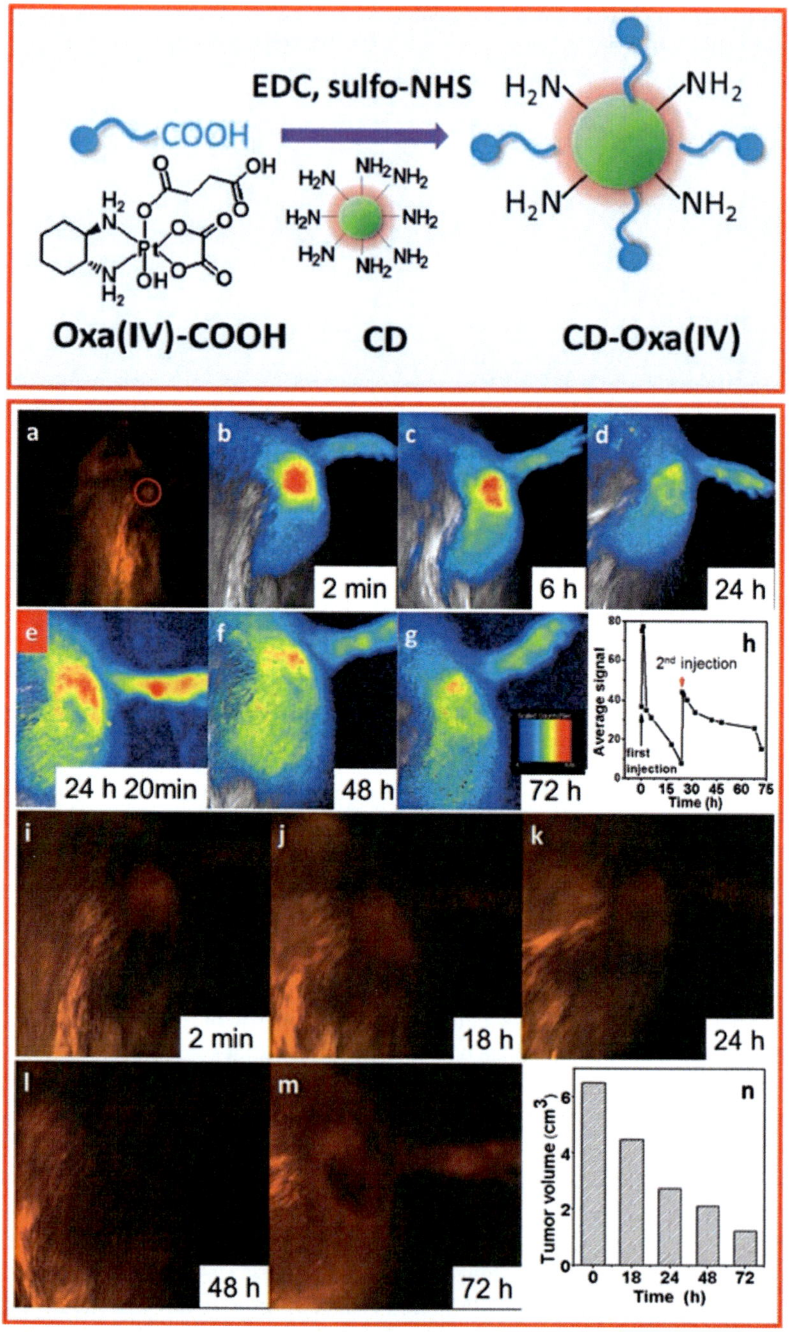
EDC, sulfo-NHS
COOH
Oxa(IV)-COOH
CD
CD-Oxa(IV)
a
b
2 min
c
6 h
d
24 h
e
24 h 20min
f
48 h
g
72 h
h
2nd injection
first injection
Average signal
Time (h)
i
2 min
j
18 h
k
24 h
l
48 h
m
72 h
n
Tumor volume (cm3)
Time (h)

◀**Fig. 5.12** Upper panel shows schematic for the synthesis of theranostic CD-Oxa complex. **a–g**, Lower panel demonstrates in vivo fluorescence imaging of CD-Oxa injected at different time periods labeled in the figure (**e**, Marks a second injection time), **h,** Semiquantitative fluorescence intensities at tumor site, **i–m,** pictographical images of tumor post-injection, **n,** the sizes of tumor measured over time. Adapted from Zheng et al. (2014)

They essentially developed a tumor microenvironment-responsive drug nanocarrier based on cisplatin (IV) prodrug-loaded charge-convertible CDs (CDs–Pt(IV)@PEG-(PAH/DMMA)) (Feng et al. 2016). The complex was developed for an imaging-guided drug delivery of cisplatin and harvesting the diagnostic potential of CDs (Fig. 5.14). Essentially, the complex was kind of a theranostic system. Briefly, a negatively charged polymer on CDs-Pt(IV) complex could undergo a switch or a charge conversion to a positively charged polymer in restricted tumor microenvironment (acidic pH $\sim$6). Thereby, such cationic nature causes high affinity binding to cell surface enhancing the internalization efficiency of the complex and locally increasing the cellular concentration of anticancer drugs. In the reducing microenvironment of cells, the prodrug is then activated causes cell cytotoxicity.

5.3 Hyperthermia-Based Therapy

One of the recent additive technologies toward intervention of cancer therapy is heat-based therapy which holds tremendous potential. Such therapies are also termed as photothermal-based therapy. Figure 5.15 shows a general schematic of how a typical photothermal-based cancer therapy is achieved using a graphene-based nanomaterial (GO, GQDs, CDs). Typically, when a laser source (near-infrared) is focused on a photo-active graphene-based nanomaterial, they absorb the energy from the light. Such absorbed energy is then converted to thermal energy thereby releasing local heat which might go several tens of celsius above room temperature depending on the laser wavelength and laser power. Such heat can cause hyperthermia in cells and tissues around it and cause cellular death. Using this principle, if the nanomaterial is in the vicinity or inside the tumor cell, it can trigger cell death. One important point to be taken into consideration is that the laser power does not cause cytotoxic effect to the other cells by itself since most of such lasers are in the near-infrared range and most cellular components do not absorb in such range.

Thus, with photothermal therapy, one can expect localized treatment of cancer by concentrating graphene-based nanomaterials in the vicinity of tumors followed by laser irradiation. It must also be noted that with photothermal therapy, one can use chemotherapeutic drugs as an additive to boost therapeutic response or completely avoid the usage of any such drug. Several research groups have explored potential synergistic effects of combined hyperthermia and chemotherapy for such improved efficacy. While others have resorted to new 'theranostic' applications, the diagnostic capability of the graphene nanomaterials is explored along with

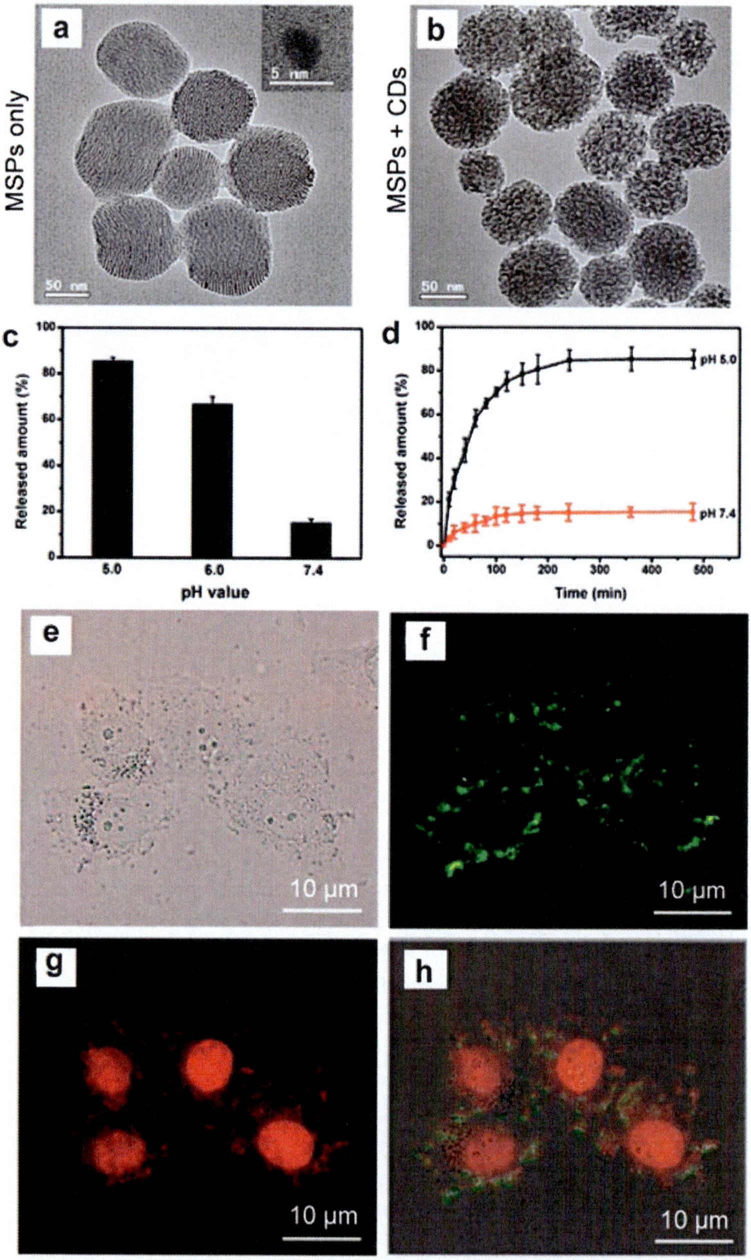
a
5 nm
MSPs only
50 nm
b
MSPs + CDs
50 nm
c
Released amount (%)
5.0
6.0
7.4
pH value
d
Released amount (%)
pH 5.0
pH 7.4
Time (min)
e
10 μm
f
10 μm
g
10 μm
h
10 μm

◂**Fig. 5.13** **a-b** TEM images of mesoporous silica (MSP) and CDs-gated MSPs, **c–d** DOX release from the complex at different pH values for 480 min, **e–h** Fluorescence images showing uptake of CDs@MSPs-DOX complex in HeLa cells. Reprinted (adapted) with permission from Zhou et al. (2013). Copyright (2013) American Chemical Society

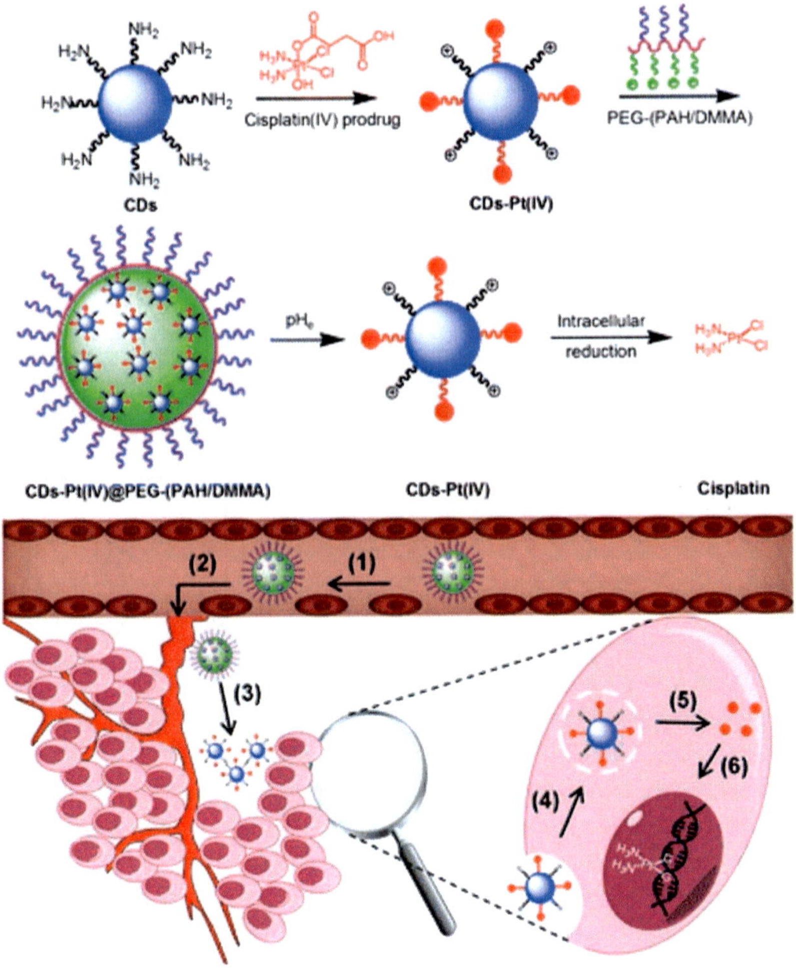

Fig. 5.14 Schematic showing preparation of charge-convertible CDs-Pt(IV)@PEG-(PAH/DMMA) complex and enhanced permeation into the tumors and entering tumor cells. Reprinted (adapted) with permission from Feng et al. (2016). Copyright (2016) American Chemical Society

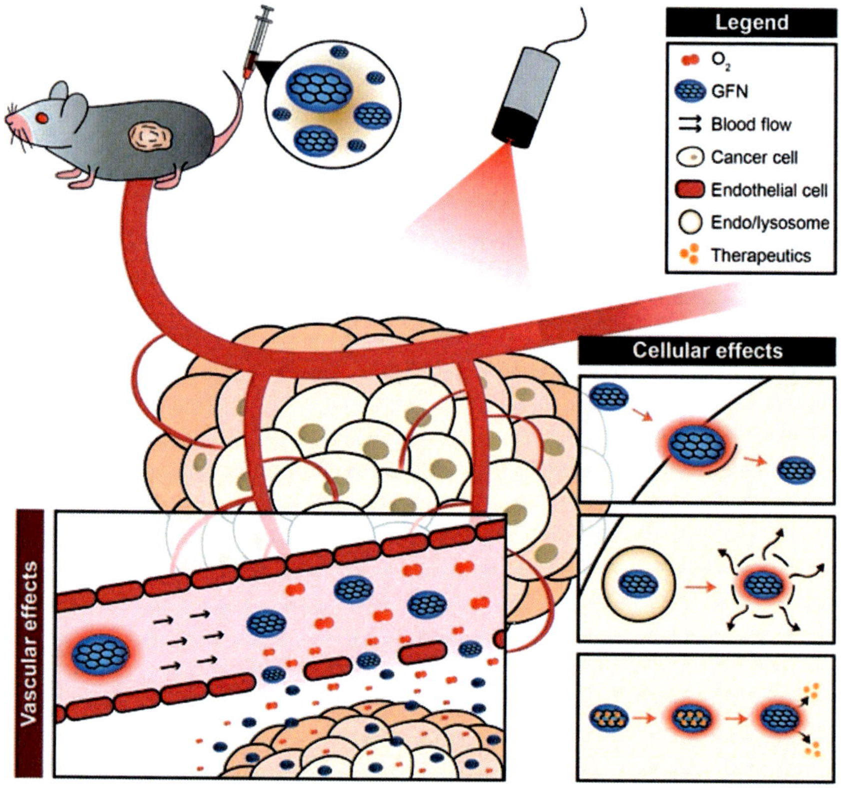

Fig. 5.15 Schematic showing a general procedure for photothermal therapy using graphene-based nanomaterials (GFN). Reproduced from Ref. De Melo-Diogo et al. (2019) with permission from The Royal Society of Chemistry

therapeutic (photothermal effect). In the following case studies, we will see many such examples and principle application of graphene-based nanomaterials for hyperthermic therapy of tumors.

In one of the earlier studies, the GQDs were synthesized from biosources such as withered leaves with hydrothermal treatment of the precursors. Subsequently, the purified GQDs were not only biocompatible but also photothermally active (Fig. 5.16a, b). The GQDs were used for photothermal therapy of breast cancer cells (Fig. 5.16c, d). The cancer cell death was ascribed to apoptosis of the cells since the temperature of the GQDs increased >42 °C. Hence with such high temperature the cells entered the apoptosis phase causing programmed cell death (Thakur et al. 2017). This was one of the first demonstration of using biosources-based GQDs for using photothermal strategy without chemotherapeutics used. The GQD were also photostable in different cellular media and after laser irradiation (Fig. 5.16e, f).

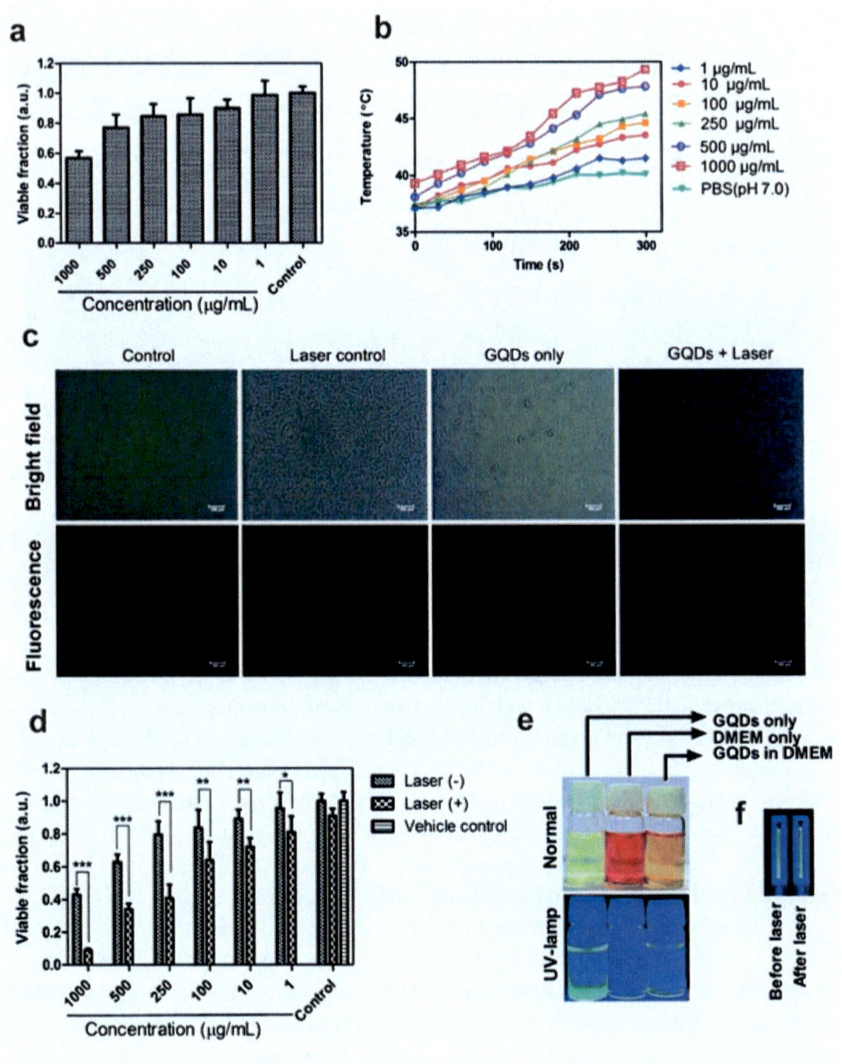

Fig. 5.16 **a** Cytotoxicity of GQDs (up to 1 mg/mL) in L929 cells, **b** hyperthermic profile showing rise in temperature after irradiation with 808 nm laser (0.5 W, 5 min), **c** cellular morphology of MDA-MB-231 cells after laser treatment for 5 min, **d** corresponding toxicity profile showing toxicity in MDA-MB-231 cells with photothermal therapy, **e** fluorescence stability of GQDs in different cellular media, **f** photostability before and after laser treatment on bare GQDs. Adapted from Thakur et al. (2017)—Published by The Royal Society of Chemistry

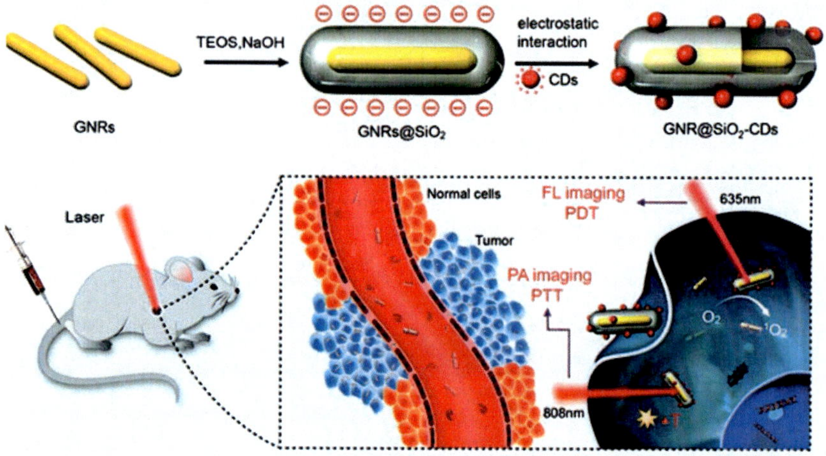

Fig. 5.17 Schematic illustration showing gold nanorods (GNR) conjugated silica-coated CDs. The complex demonstrates as a theranostic agent for cancer therapy. Adapted from Jia et al. (2016)—Published by The Royal Society of Chemistry

Modification and more research work involving using CDs as an additional asset toward synergistic modality has been carried out recently. For instance, gold nanorods (GNRs) were coated with silica-CDs using silica as a scaffold (Fig. 5.17). The complex was not truly a multifunctional. The GNRs were capable of photothermal and photoacoustic response while CDs would serve as imaging agent. In a similar manner, GQDs were decorated on GO to form GO-GQD complexes (Fig. 5.18). The GO-GQD complex were formed electrostatic interaction for enhanced photothermal therapy of cancer (Kumawat et al. 2019).

Higher efficiency absorbance and effective photothermal therapy are ideally required in photothermal therapy of cancers. The CDs with longer wavelength absorbance, near infrared wavelength (NIR-II) window which correspond to 1000–1700 nm were developed by Li and co-workers (Li et al. 2019). Interestingly, the CDs were developed using watermelon juice as a precursor source. A high quantum yield CDs were developed with an absorbance with 900–1200 nm. They demonstrated the sue of CDs for NIR imaging probes and simultaneously photothermal therapeutics in mouse models (Fig. 5.19).

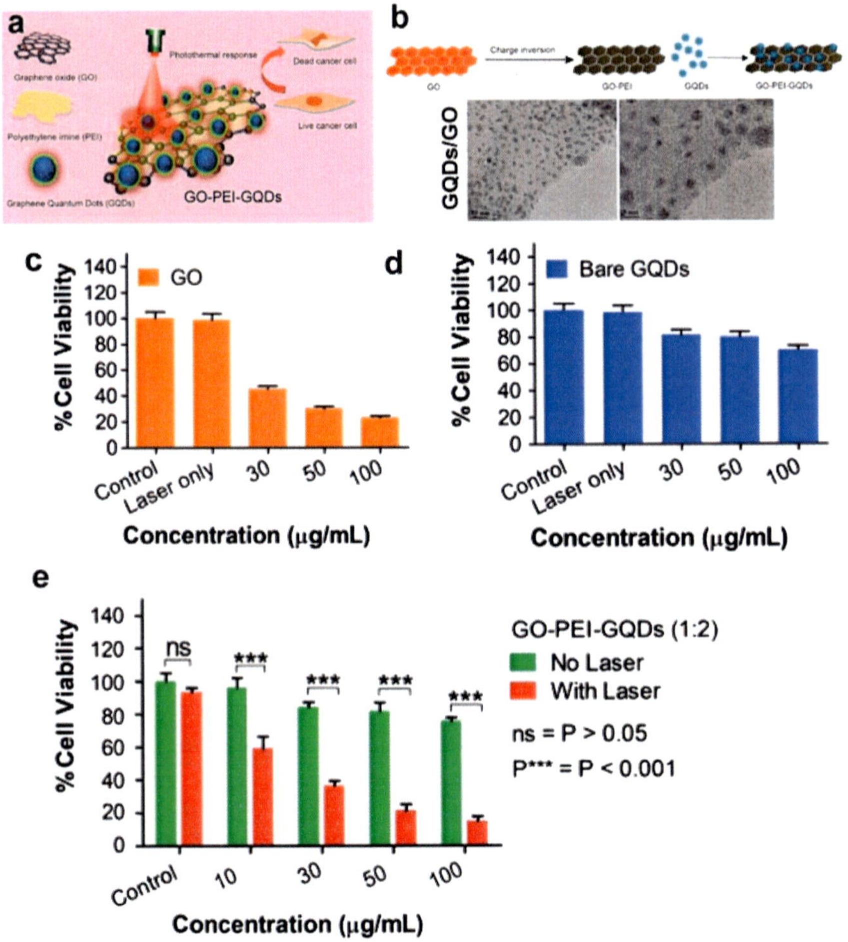

Fig. 5.18 **a** Photothermal therapy of cancer cells using GO-GQD complex, **b** synthesis of GO-PEI-GQDs, **c** cell viability analysis of GO in MDA-MB-231 cells, **d** cell viability analysis of GQDs in MDA-MB-231 cells, e photothermal treatment of MDA-MB-231 cells with GO-PEI-GQDs complex with 808 nm laser (0.5 W). Adapted from Kumawat et al. (2019)

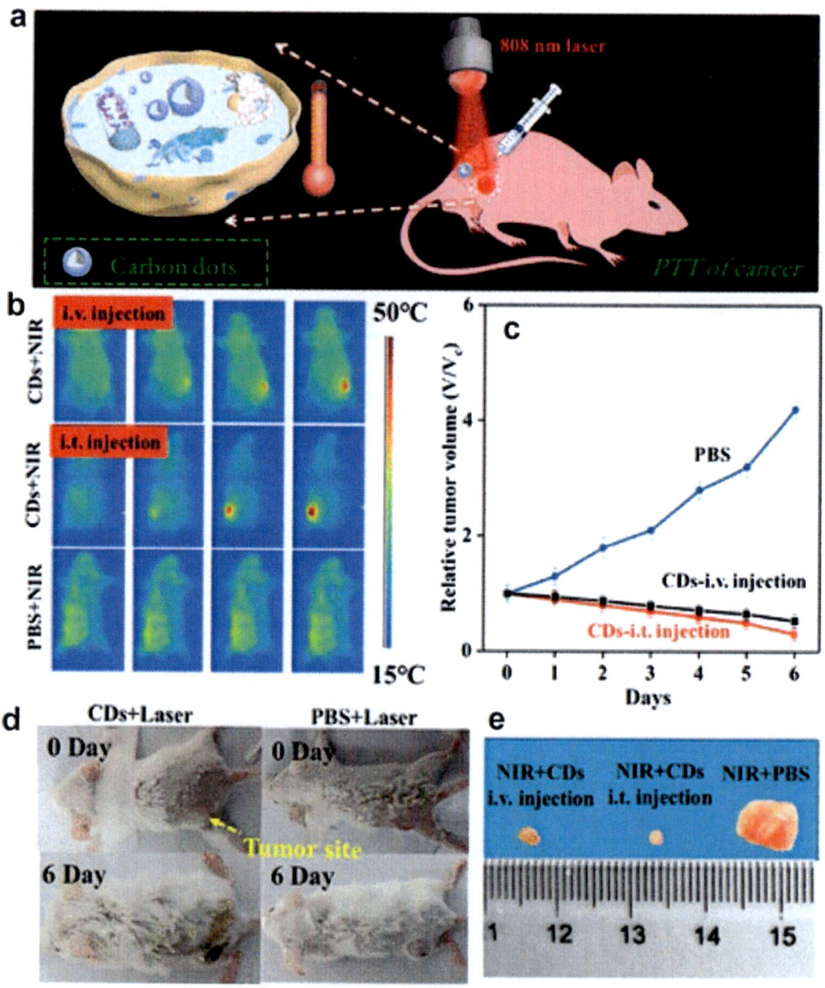

Fig. 5.19 **a** Synthesis of NIR-II absorbing CDs and their use as photothermal agent, **b** infrared images showing heat generation in mice injected with CDs via intravenous (i.v.) or intratumoral (i. t.), **c** measured relative tumor volume followed up after treatment with CDs, **d** pictorial images of mice showing shrinkage of tumors, **e** measured tumor volume after photothermal therapy. Reprinted (adapted) with permission from Li et al. (2019). Copyright (2019) American Chemical Society

5.4 Photodynamic Therapy

Photodynamic therapy refers to the production of singlet oxygen species ($^1O_2$) which can cause oxidative damage to cancer cells. Such singlet oxygen species are a major component of reactive oxygen species (ROS) that are major causes of

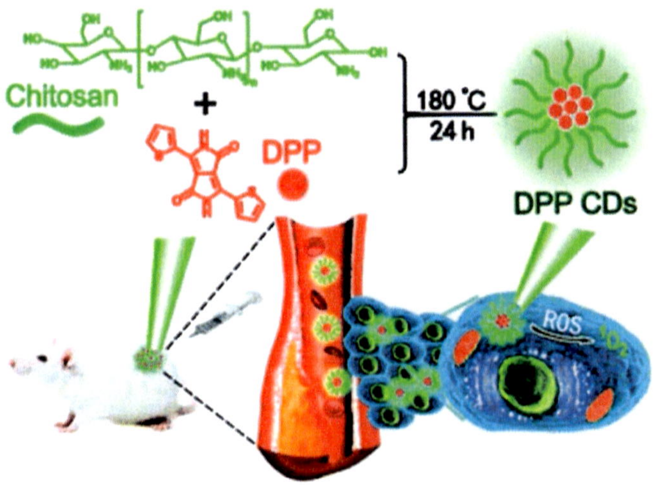

Fig. 5.20 Schematic showing fluorescent CDs synthesis and its photodynamic effect in cancer mouse models. Adapted from He et al. (2018)—Published by The Royal Society of Chemistry

oxidative stress in cells. A laser power is irradiated on a photosensitizer which absorbs specific wavelength ranges thereby producing ROS species predominantly singlet oxygen species. Recently, graphene-based nanomaterials have also been shown to have photodynamic effect in tumor cells. This is due to their biocompatibility and high photosensitizing property. In the following examples, we will see different graphene-based nanomaterials mostly CDs and GQDs being used as photodynamic agents along with their other properties in the field of cancer therapy.

Figure 5.20 shows a typical synthesis of CDs from diketopyrrolopyrrole (DPP)—a type of light absorbing pigment with exceptional optical properties. CDs synthesized using DPP were able to generate $^1O_2$ in vitro HepG2 cells. The studies in mouse models also showed that the tumor volume decreased dramatically in 12 days upon irradiation with 540 nm laser. The key finding of this study was that a new therapeutic modality was demonstrated in the form of oxidative stress mediated cellular death of cancer cells using a non-invasive technique. Recently, the photodynamic therapy and mechanism of cell toxicity was studied in glioma cells by Markovic and co-workers (Markovic et al. 2012). They showed that the GQDs incubated in glioma cells when irradiated with a blue light (470 nm) the GQD is photoactivated generating singlet oxygen species. The cell undergoes morphological changes upon activation of the GQDs and then enters either apoptosis or autophagy (Fig. 5.21). The studies explored the photodynamic effect of GQDs but also highlight the concerns over the possible inherent toxicity of the nanoparticles. Few other approaches exploring photodynamic therapy have been using rare earth

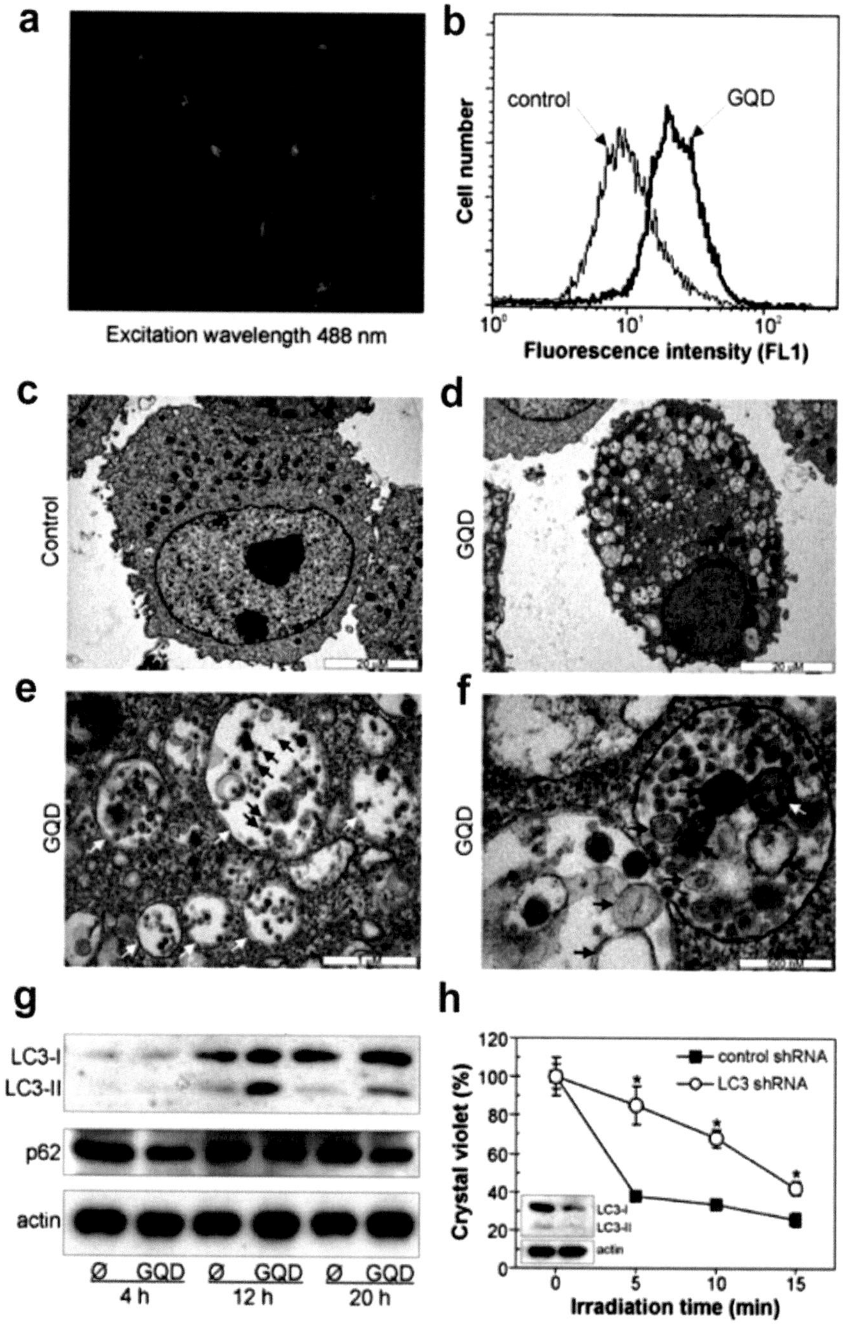
a
Excitation wavelength 488 nm
b
control
GQD
Cell number
Fluorescence intensity (FL1)
c
Control
d
GQD
e
GQD
f
GQD
g
LC3-I
LC3-II
p62
actin
Ø GQD Ø GQD Ø GQD
4 h 12 h 20 h
h
control shRNA
LC3 shRNA
Crystal violet (%)
Irradiation time (min)
LC3-I
LC3-II
actin

◂**Fig. 5.21** Autophagy and mechanism of toxicity of GQDs in cells. **a, b** GQDs treated (0.2 mg/mL) U251 cells and irradiated with 470 nm laser (1 W), **c, d** TEM imaging of U251 cells incubated without or with GQDs, **e** GQDs in intracellular vesicles, **f** GQDs in autophagic vacuoles, **g** immunoblotting staining showing levels of LC3 conversion and p62 levels in U251 cells post-incubation with GQDs, **h** U251 cells transfected using control or LC3B shRNA and treated with GQDs. Adapted from Markovic et al. (2012)

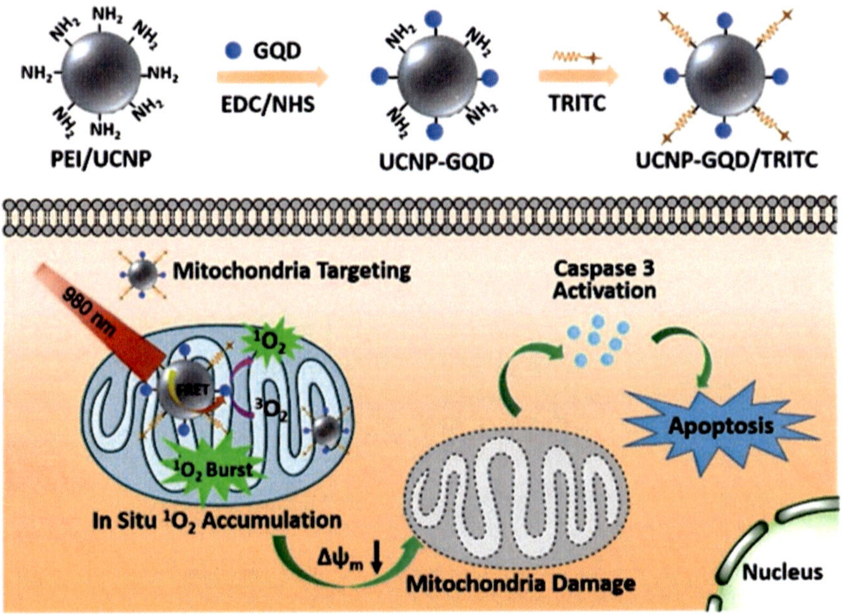

Fig. 5.22 Schematic showing synthesis of upconversion nanoparticle with GQDs and mitochondrial targeting rhodamine derivative (TRITC). The complex is photodynamically active causing ROS production in cancer cells. Adapted from (Zhang et al. 2018)

elements and GQDs conjugation for targeting specific organelles in cells (Zhang et al. 2018). In the search for better and more efficient photodynamic sensitizers, Zhang and co-workers synthesized GQDs with UCNPs doped with rare earth metals (Fig. 5.22). The resultant complex could absorb light from longer wavelength and emit radiation in shorter wavelength (visible light spectrum). Thus, the GQDs were also able to generate ROS more efficiently. Using a rhodamine derivative TRITC, they could target mitochondria in cells and cause changes in the mitochondria membrane potential. The reaction would then undergo irreversible cell apoptosis eventually killing the cancer cell. Thakur and co-workers also analyzed the singlet oxygen species production in cells using bare GQDs synthesized using watered

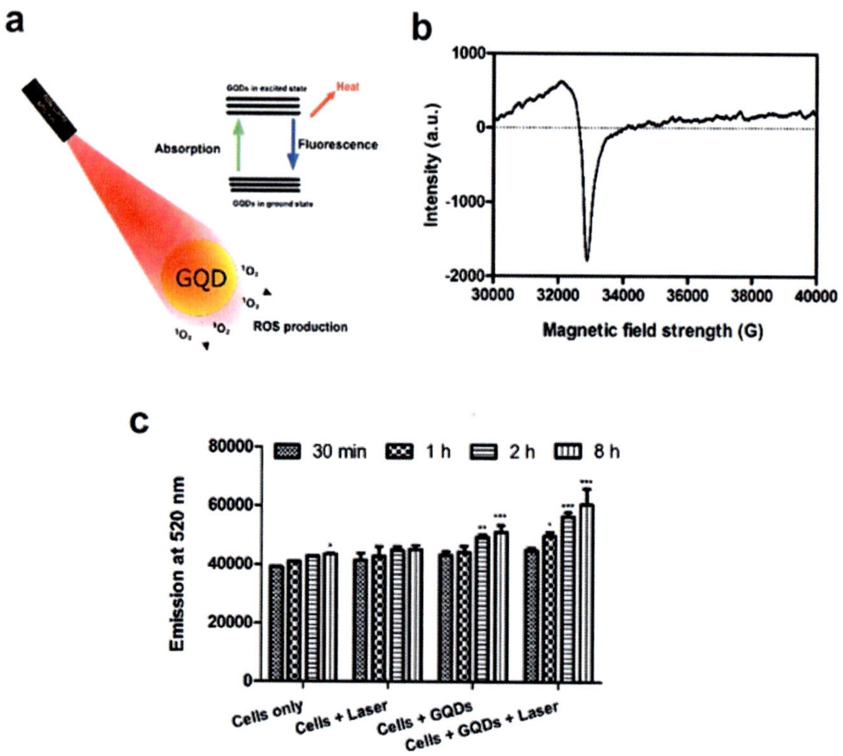

Fig. 5.23 **a** Mechanism of combined photothermal and photodynamic effect in GQDs with a NIR laser, **b** Electron spin resonance (ESR) showing ROS production, **c** Fluorescence of ROS sensitive dye (DCFDA) showing higher ROS production upon laser irradiation in MDA-MB-231 cell lines. Adapted from Thakur et al. (2017)—Published by The Royal Society of Chemistry

leaves (Thakur et al. 2017). Figure 5.23 shows the generation of singlet oxygen species as observed using ESR spectrum. Using photothermal and photodynamic therapy, one could enhance cancer therapeutics.

5.5 Biosensing of Cancer Biomarkers

Several biomarkers of cancer have been used as a benchmark to target or to detect in cancer diagnostics. Graphene-based nanomaterials for cancer biomarkers detection such as specific overexpressed cell receptor, over-expressed gene/proteins, and even high pH/temperature could be used for rapid diagnosis of cancer. Recently, CDs have been extensively research for such application as shown in Fig. 5.24.

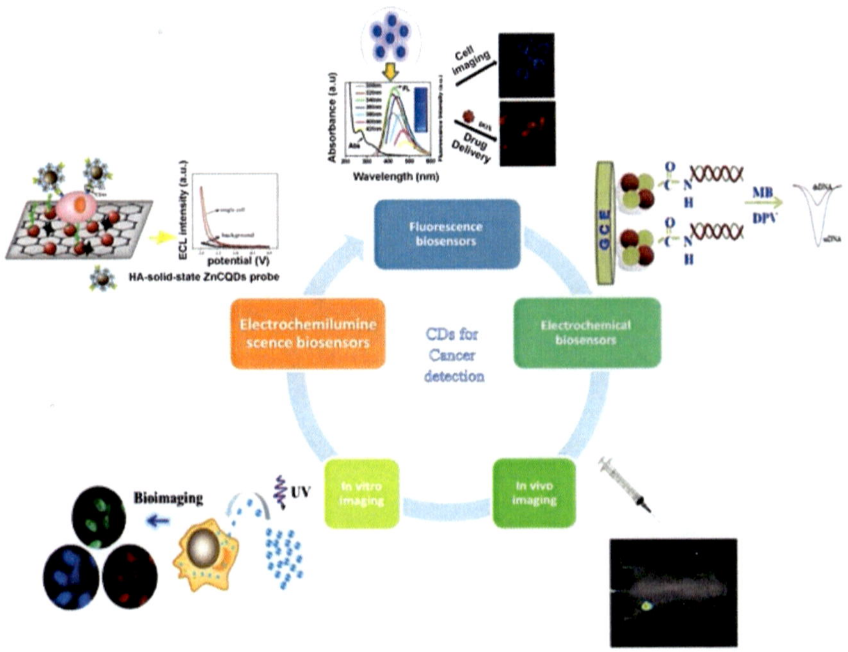

Fig. 5.24 Schematic showing different application of CDs in cancer biomarker detection. Adapted from (Pirsaheb et al. 2019)

Liu and co-workers developed a simple strategy to detect cancer cells by using ON/OFF fluorescence scheme using CDs (Fig. 5.25). They synthesized CDs and decorated with folic acid (FA). The cancer cells often overexpressed with folic acid receptors were labeled using FA-CDs complex in a turn ON strategy. The FA-CDs could easily differentiate FA-positive cells from others. In the absence of such folic acid receptors, the system does not show fluorescence. This study has an important application in live surgical treatment of cancer where a labeled cell could help the surgeons to selectively remove the cancer tissue from the normal ones.

Alpha fetal protein (AFP) is an important tumor marker for diagnosing hepatocellular cancer (AFP level > 25 ng/mL in serum is associated with the liver cancer) (Xu et al. 2017). Hence, two methods based on immunohybridization chain reaction (immuno-HCR) and metal-enhanced fluorescence (MEF) of CDs were used for the detection of alpha fetal protein (AFP). In such strategy (Fig. 5.26), the plasmonic surface antibody captures AFP and then a secondary antibody with an oligonucleotide initiator is added that binds to AFP targets on the slide. Further, CDs labeled DNA hairpins are added that triggers HCR thereby forming complex. The detection is using CDs is enhanced using metal-induced fluorescent enhancement, this technique is very specific and shows the application of CDs in synergy with plasmonic applications. Therapy and live cell tracking is another important application of fluorescent probes for cancer cell diagnosis and surgery.

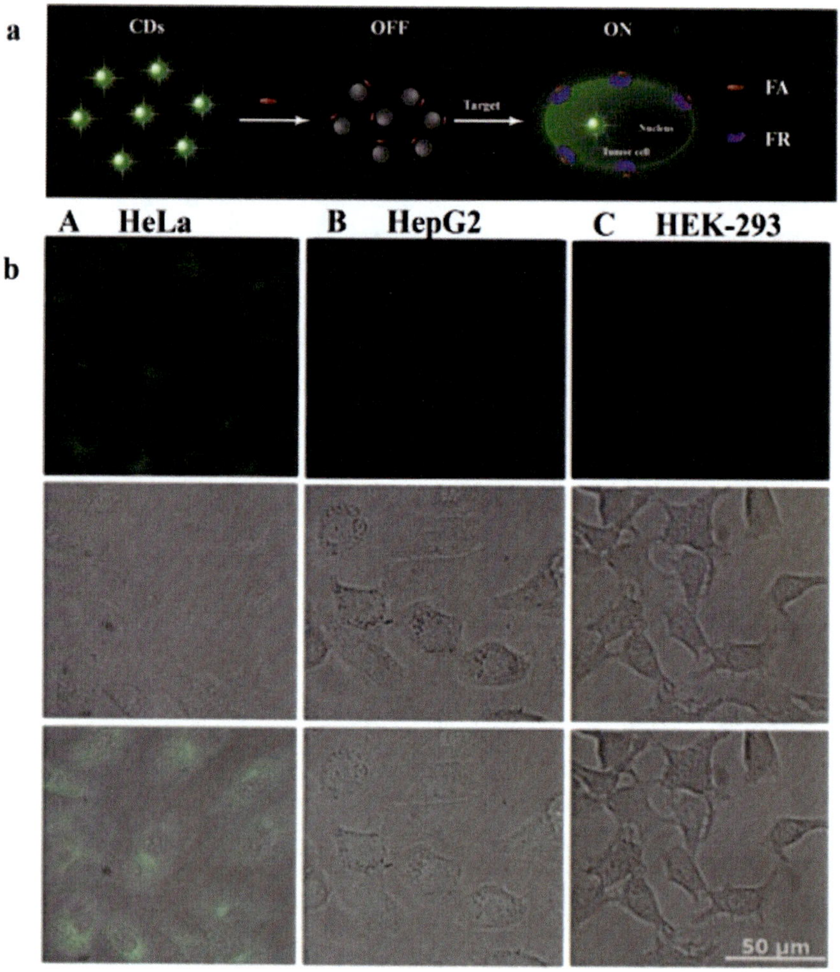

Fig. 5.25 **a** Schematic showing turn ON/OFF folic acid receptor-based biosensor using FA-CDs, **b** Cells incubated with FA-CDs (2 h) show turn on fluorescence in cancer cells. Adapted from Liu et al. (2015)

A dual-strategy was used to photothermally treat MDA-MB-231 cells along with simultaneous cell tracking. Figure 5.27 shows the cell tracing application of GQDs up to 24 h where the fluorescence of the GQDs can be seen decreasing with time as the cells die out of photothermal and photodynamic effect. As the cell dies, the plasma membrane bursts open and the GQDs are removed from the cells. Thus, such dual-strategy can be used for simultaneous therapy and tracking of the cells.

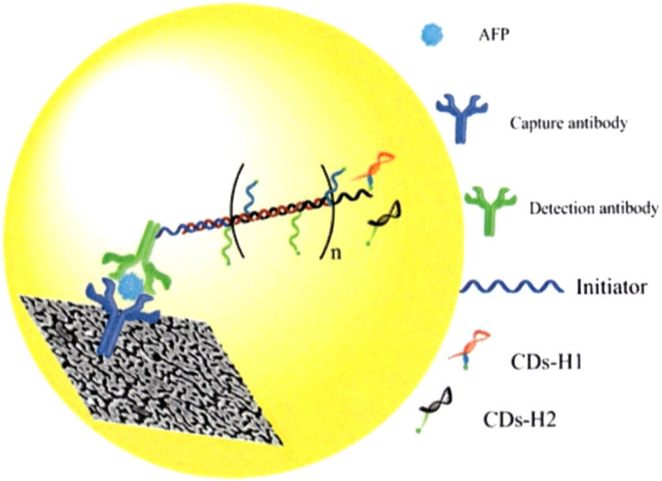

Fig. 5.26 **a** Schematic showing detection of alpha fetal protein (AFP) using fluorescence based on CDs. Adapted from Xu et al. (2017)

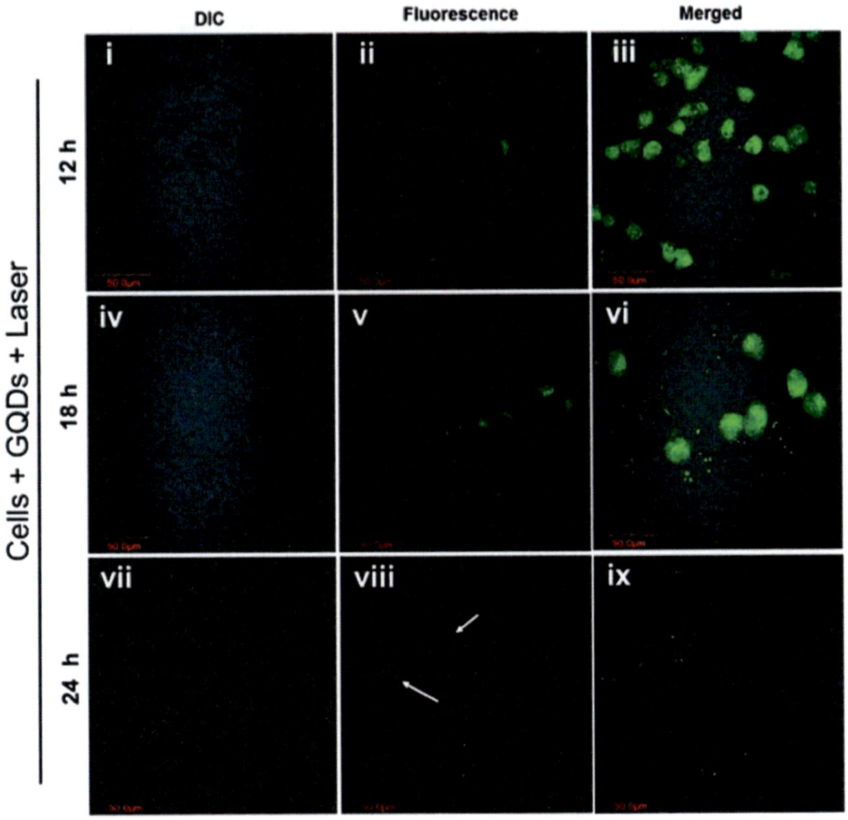

Fig. 5.27 Live cell tracking of cancer cells using GQDs-tagged MDA-MB-231 cells using by time-lapsed confocal imaging in vitro. Adapted from (Thakur et al. 2017)

References

Chen, B. Bin, et al. (2019). Fluorescent carbon dots functionalization. *Advances in Colloid and Interface Science*, 165–190. https://doi.org/10.1016/j.cis.2019.06.008. (Elsevier B.V.).

Chen, F., et al. (2017). Graphene quantum dots in biomedical applications: Recent advances and future challenges. *Frontiers in Laboratory Medicine, 1*(4), 192–199. https://doi.org/10.1016/J.FLM.2017.12.006. (Elsevier).

De Melo-Diogo, D., et al. (2019). Graphene family nanomaterials for application in cancer combination photothermal therapy. *Biomaterials Science*, 3534–3551. https://doi.org/10.1039/c9bm00577c. (Royal Society of Chemistry).

Feng, T., et al. (2016). Charge-convertible carbon dots for imaging-guided drug delivery with enhanced in vivo cancer therapeutic efficiency. *ACS Nano, 10*(4), 4410–4420. https://doi.org/10.1021/acsnano.6b00043. (American Chemical Society).

Geim, A. K., & Novoselov, K. S. (2007). The rise of graphene. *Nature Materials, 6*(3), 183–191. https://doi.org/10.1038/nmat1849. (Nature Publishing Group).

He, H., et al. (2018). Diketopyrrolopyrrole-based carbon dots for photodynamic therapy. *Nanoscale, 10*(23), 10991–10998. https://doi.org/10.1039/c8nr02643b. (Royal Society of Chemistry).

Hong, G., et al. (2015). Carbon nanomaterials for biological imaging and nanomedicinal therapy. *Chemical Reviews*, 10816–10906. https://doi.org/10.1021/acs.chemrev.5b00008. (American Chemical Society).

Jia, Q., et al. (2016). Gold nanorod@silica-carbon dots as multifunctional phototheranostics for fluorescence and photoacoustic imaging-guided synergistic photodynamic/photothermal therapy. *Nanoscale, 8*(26), 13067–13077. https://doi.org/10.1039/c6nr03459d. (Royal Society of Chemistry).

Khodadadei, F., Safarian, S., & Ghanbari, N. (2017). Methotrexate-loaded nitrogen-doped graphene quantum dots nanocarriers as an efficient anticancer drug delivery system. *Materials Science and Engineering C, 79,* 280–285. https://doi.org/10.1016/j.msec.2017.05.049. (Elsevier Ltd).

Kong, T., et al. (2018). Doxorubicin conjugated carbon dots as a drug delivery system for human breast cancer therapy. *Cell Proliferation, 51*(5), e12488. https://doi.org/10.1111/cpr.12488. (Blackwell Publishing Ltd).

Kumawat, M. K., et al. (2019). Preparation of graphene oxide-graphene quantum dots hybrid and its application in cancer theranostics. *Materials Science and Engineering C, 103,* 109774. https://doi.org/10.1016/j.msec.2019.109774. (Elsevier Ltd).

Li, Y., et al. (2019). Theranostic carbon dots with innovative NIR-II emission for in vivo renal-excreted optical imaging and photothermal therapy. *ACS Applied Materials and Interfaces, 11*(5), 4737–4744. https://doi.org/10.1021/acsami.8b14877. (American Chemical Society).

Liang, J., et al. (2019). pH-responsive nanoparticles loaded with graphene quantum dots and doxorubicin for intracellular imaging, drug delivery and efficient cancer therapy. *Chemistry Select, 4*(20), 6004–6012. https://doi.org/10.1002/slct.201803807. (Wiley-Blackwell).

Liu, Z., et al. (2011). Carbon materials for drug delivery & cancer therapy. *Materials Today, 14*(7–8), 316–323. https://doi.org/10.1016/s1369-7021(11)70161-4. (Elsevier B.V.).

Liu, Q., et al. (2015). Distinguish cancer cells based on targeting turn-on fluorescence imaging by folate functionalized green emitting carbon dots. *Biosensors and Bioelectronics, 64,* 119–125. https://doi.org/10.1016/j.bios.2014.08.052. (Elsevier Ltd).

Markovic, Z. M., et al. (2012). Graphene quantum dots as autophagy-inducing photodynamic agents. *Biomaterials, 33*(29), 7084–7092. https://doi.org/10.1016/j.biomaterials.2012.06.060. (Elsevier).

Mohajeri, M., Behnam, B., & Sahebkar, A. (2018). Biomedical applications of carbon nanomaterials: Drug and gene delivery potentials. *Journal of Cellular Physiology*, 298–319. https://doi.org/10.1002/jcp.26899. (Wiley-Liss Inc.).

Novoselov, K. S., et al. (2004). Electric field in atomically thin carbon films. *Science, 306*(5696), 666–669. https://doi.org/10.1126/science.1102896. (American Association for the Advancement of Science).

Pirsaheb, M., Mohammadi, S., & Salimi, A. (2019). Current advances of carbon dots based biosensors for tumor marker detection, cancer cells analysis and bioimaging. *TrAC—Trends in Analytical Chemistry*, 83–99. https://doi.org/10.1016/j.trac.2019.04.003. (Elsevier B.V.).

Sun, T., et al. (2014). Engineered nanoparticles for drug delivery in cancer therapy. *Angewandte Chemie—International Edition*, 12320–12364. https://doi.org/10.1002/anie.201403036. (Wiley-VCH Verlag).

Thakur, M., Kumawat, M. K., & Srivastava, R. (2017). Multifunctional graphene quantum dots for combined photothermal and photodynamic therapy coupled with cancer cell tracking applications. *RSC Advances, 7*(9), 5251–5261. https://doi.org/10.1039/c6ra25976f. (Royal Society of Chemistry).

Tian, P., et al. (2018). Graphene quantum dots from chemistry to applications. *Materials Today Chemistry, 10,* 221–258. https://doi.org/10.1016/J.MTCHEM.2018.09.007. Elsevier.

Wang, Q., et al. (2013). Hollow luminescent carbon dots for drug delivery. *Carbon, 59,* 192–199. https://doi.org/10.1016/j.carbon.2013.03.009. (Pergamon).

Wang, Q., et al. (2014a). Fluorescent carbon dots as an efficient siRNA nanocarrier for its interference therapy in gastric cancer cells. *Journal of Nanobiotechnology, 12*(1), 58. https://doi.org/10.1186/s12951-014-0058-0. (BioMed Central Ltd.).

Wang, X., et al. (2014b). Multifunctional graphene quantum dots for simultaneous targeted cellular imaging and drug delivery. *Colloids and Surfaces B: Biointerfaces, 122,* 638–644. https://doi.org/10.1016/j.colsurfb.2014.07.043. (Elsevier).

Wang, F., et al. (2019). Conjugated polymer dots for biocompatible siRNA delivery. *New Journal of Chemistry, 43*(36), 14443–14449. https://doi.org/10.1039/c9nj03277k. (Royal Society of Chemistry).

Wu, Y. F., et al. (2016). Multi-functionalized carbon dots as theranostic nanoagent for gene delivery in lung cancer therapy. *Scientific Reports, 6*(1), 1–12. https://doi.org/10.1038/srep21170. (Nature Publishing Group).

Xie, J., Lee, S., & Chen, X. (2010). Nanoparticle-based theranostic agents. *Advanced Drug Delivery Reviews*, 1064–1079. https://doi.org/10.1016/j.addr.2010.07.009. (Elsevier).

Xu, D. D., et al. (2017). Dual amplification fluorescence assay for alpha fetal protein utilizing immunohybridization chain reaction and metal-enhanced fluorescence of carbon nanodots. *ACS Applied Materials and Interfaces, 9*(43), 37606–37614. https://doi.org/10.1021/acsami.7b11659. (American Chemical Society).

Yan, Y., et al. (2019). Recent advances on graphene quantum dots: from chemistry and physics to applications. *Advanced Materials, 31*(21), 1808283. https://doi.org/10.1002/adma.201808283. (John Wiley & Sons Ltd).

Zhang, J., Zheng, M., & Xie, Z. (2016). Co-assembled hybrids of proteins and carbon dots for intracellular protein delivery. *Journal of Materials Chemistry B, 4*(34), 5659–5663. https://doi.org/10.1039/c6tb01622g. (Royal Society of Chemistry).

Zhang, D., et al. (2018). Mitochondrial specific photodynamic therapy by rare-earth nanoparticles mediated near-infrared graphene quantum dots. *Biomaterials, 153,* 14–26. https://doi.org/10.1016/j.biomaterials.2017.10.034. (Elsevier Ltd).

Zheng, X. T., et al. (2015). Glowing graphene quantum dots and carbon dots: Properties, syntheses, and biological applications. *Small*, 1620–1636. https://doi.org/10.1002/smll.201402648. (Wiley-VCH Verlag).

Zheng, M., et al. (2014). Integrating oxaliplatin with highly luminescent carbon dots: An unprecedented theranostic agent for personalized medicine. *Advanced Materials, 26*(21), 3554–3560. https://doi.org/10.1002/adma.201306192. (Wiley-VCH Verlag).

Zhou, L., et al. (2013). Luminescent carbon dot-gated nanovehicles for ph-triggered intracellular controlled release and imaging. *Langmuir, 29*(21), 6396–6403. https://doi.org/10.1021/la400479n. (American Chemical Society).

Chapter 6
Outlook, Challenges, and Future Perspectives

In this book, we discussed the historical aspects of carbon allotropes such as graphene oxide and associated new generation graphene-based nanomaterials such as graphene quantum dots/carbon dots, and so on. The discovery of such 'graphene-based' nanomaterials led to thousands of research publications exploring the multitude of properties possessed by them. Prominently, we discussed their exciting physicochemical properties and their role in theranostic applications, in the context of diagnosis and therapy of cancer. We also discussed their possible challenges being faced currently in cancer therapy and provide insights into various functionalization techniques carried out with graphene nanomaterials, to enhance their efficacy and applicability in cancer.

New generation graphene-based nanomaterials' current and potential bioapplications are astounding. In cancer theranostics, one of the essential properties which is required for these nanomaterials is the aqueous dispersibility which is a challenge in graphene-based nanomaterials. This was seen to be overcome by modifications with various oxygen and nitrogen functional groups. Although the industrial sector and academic researchers are showing great enthusiasm to revolutionize graphene-based products, especially in the field of medicine, they are still in a phase of research and development and have to face numerous barriers before establishing itself into the present market. The need of the hour is to power more academic research, so that the full capabilities could be discovered and how it can be translated into real-life applications. The production cost of such graphene-based nanomaterials poses another challenge toward commercialization. For instance, to produce a defect-free monolayer of graphene, exfoliation is still the best technique to date. This technique is advantageous to create, evolve, and study graphene in academic research but is not suitable for high-throughput large-scale production. Nevertheless, the available state of the art is not fully ready for industrial applications, and new innovative approaches are actively sought. Although this example

R. Srivastava et al., *Next Generation Graphene Nanomaterials for Cancer Theranostic Applications*,
https://doi.org/10.1007/978-981-33-6303-8_6

holds true for 2D graphene, for other next-generation nanomaterials there are new challenges in the pipeline. The three major reasons are the quality of the product in terms of production reproducibility, purity factor (or yield), and limited clinical studies demonstrating the applicability of these nanomaterials.

6.1 Reproducibility of the Product

The synthetic methods involved in the production of graphene-based nanomaterials (e.g., GQDs) comprise either top-down or bottom-up approaches. Although these approaches are facile and convenient, both these approaches are limited by the reproducibility of the final product. Most such processes include high thermal annealing processes such as microwave heating, pyrolysis, and carbonization. Owing to the difficulty in the process control, every synthesis process leads to slight variation in the nanomaterial yield reflecting the quality of the final product. Hence, the challenge is to develop reliable synthesis processes in order to get reproducible products. The amount of initial carbon-based raw material required is much higher than the actual produced graphene-based nanomaterial of interest, decreasing the production yield. Furthermore, the synthesis is also affected by initial precursor (elemental content), reaction temperature, time and even pH, as reviewed here (Luo et al. 2013; Tian et al. 2018; Xia et al. 2019).

6.2 Broad Distribution of Nanomaterials

This issue refers to the synthesis approaches used for the graphene-based nanomaterials. In a sense, every process produces a broad range of nanomaterials be it in size, shape, or even structures. For example, the GQDs produced from a green-synthesis carbonization process of complex precursor—leaves as a carbon source could have multitude of carbonaceous contents that lead to various nanomaterials post-synthesis (Figs. 6.1 and 6.2) (Thakur et al. 2017). Such processes often involve further separation steps such as dialysis and centrifugation being most widely used ones (Xia et al. 2019). Such additional steps, in most cases, lower down the actual yield and quality of the product and further increase the production time. Thus, one of the challenges is to develop synthesis processes that produce a smaller number of such broad distribution of nanomaterials.

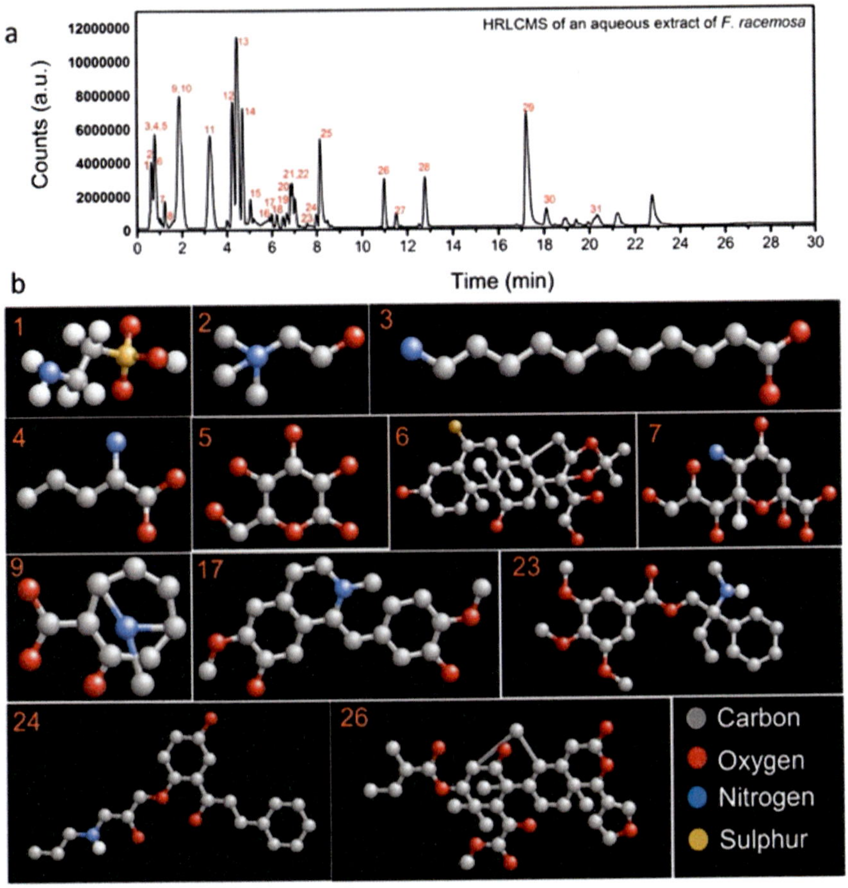

Fig. 6.1 **a** Liquid chromatography mass spectrometry of an aqueous extract of *F. racemosa* leaves. **b** Chemical structures of the compounds extracted from such an aqueous extract. Adapted from Thakur et al. (2017)

6.3 Limited Clinical Trial Studies

The excellent biocompatibility, dispersibility, and ability to cargo drugs make graphene-based nanocarriers an excellent choice for clinical studies. Furthermore, inherent optical properties make them advantageous for theranostic application in cancer therapy. There are limited studies showing the use of graphene-based nanomaterials such as CDs, GQDs, or polymer dots. Primary reason being the quality and reproducibility in the synthesis of such nanomaterials is very challenging. Furthermore, variability in terms of size, shape, and quantum yield further limits its clinical applications.

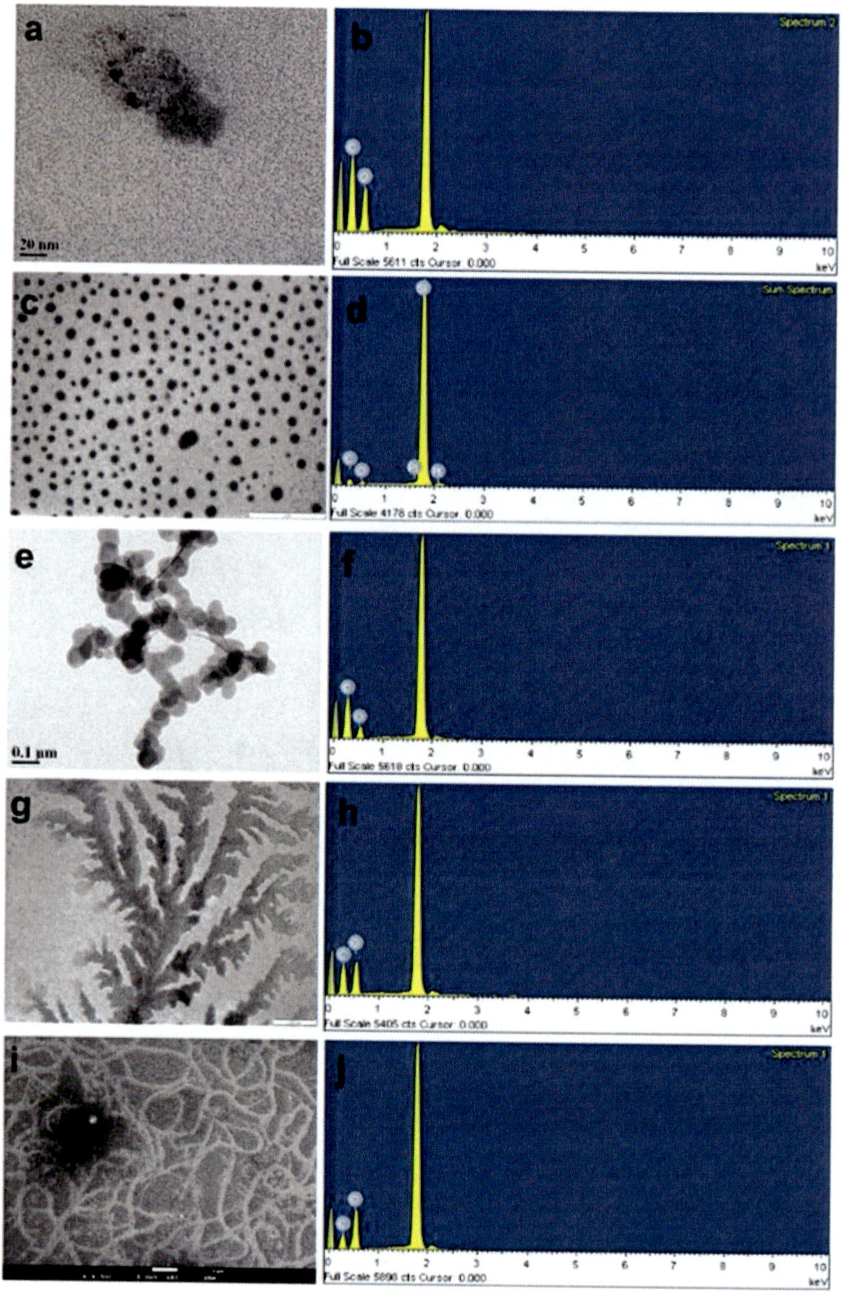

Fig. 6.2 Different carbon nanomaterials formed during synthesis of GQDs using F. racemosa leaves. Left panel is electron microscope images of different carbon-based structures (left panel) and their corresponding elemental analysis (right panel). Adapted from Thakur et al. (2017)

6.4 Commercialization

All these factors affect the successful commercialization of the graphene-based products is the safety of the concerned production technologies. Most of the methods are considered non-viable due to many potentially hazardous chemicals and substances involved in the process. The specific requirement of the synthesis like inert surrounding, vacuum containers make the synthesis more costly and tedious. Different methods of synthesizing graphene-based nanomaterials have been reviewed to date to list few (Li et al. 2019; Yan et al. 2019). The physical, chemical, mechanical, and electrical properties of graphene are highly sensitive to the preparation parameters applied and therefore it is important to understand the suitability of the particular method for the application in question. The lack of a universal method of synthesis is one of the various problems standing in the way of the successful commercialization of graphene. The graphene-based nanomaterials such as GQDs have proven to be an excellent candidate in the field of nanomedicine as a fluorophore, photothermal agent, biosensors, and for drug delivery (Liu et al. 2011; Wang et al. 2014; Hong et al. 2015; Wu et al. 2016; Chen et al. 2017; Thakur et al. 2017; Yan et al. 2019). It has emerged as one of the finest NIR-II fluorescence imaging agents due to the reduced scattering of photons which increases penetration depths. This helps in overcoming the challenge faced by traditional fluorescence imaging to visualize deep features up to a few millimeters to centimeters in a living organism.

References

Chen, F., et al. (2017). Graphene quantum dots in biomedical applications: Recent advances and future challenges. *Frontiers in Laboratory Medicine, 1*(4), 192–199. https://doi.org/10.1016/J.FLM.2017.12.006. (Elsevier).

Hong, G. et al., (2015). Carbon nanomaterials for biological imaging and nanomedicinal therapy. *Chemical Reviews*, 10816–10906. https://doi.org/10.1021/acs.chemrev.5b00008. (American Chemical Society).

Li, Y., et al. (2019). Theranostic carbon dots with innovative NIR-II emission for in vivo renal-excreted optical imaging and photothermal therapy. *ACS Applied Materials & Interfaces, 11*(5), 4737–4744. https://doi.org/10.1021/acsami.8b14877. (American Chemical Society).

Liu, Z. et al., (2011). Carbon materials for drug delivery & cancer therapy. Materials Today, *14*(7–8), 316–323. https://doi.org/10.1016/s1369-7021(11)70161-4. (Elsevier B.V.).

Luo, P. G., et al. (2013). Carbon "quantum" dots for optical bioimaging. *Journal of Materials Chemistry B, 1*(16), 2116–2127. https://doi.org/10.1039/c3tb00018d.

Thakur, M., Kumawat, M. K., & Srivastava, R. (2017). Multifunctional graphene quantum dots for combined photothermal and photodynamic therapy coupled with cancer cell tracking applications. *RSC Advances, 7*(9), 5251–5261. https://doi.org/10.1039/c6ra25976f. (Royal Society of Chemistry).

Tian, P., et al. (2018). Graphene quantum dots from chemistry to application. *Materials Today Chemistry, 10,* 221–258. https://doi.org/10.1016/J.MTCHEM.2018.09.007. (Elsevier).

Wang, X., et al. (2014). Multifunctional graphene quantum dots for simultaneous targeted cellular imaging and drug delivery. *Colloids and Surfaces B: Biointerfaces, 122,* 638–644. https://doi.org/10.1016/j.colsurfb.2014.07.043. (Elsevier).

Wu, Y. F., et al. (2016). Multi-functionalized carbon dots as theranosticnanoagent for gene delivery in lung cancer therapy. *Scientific Reports, 6*(1), 1–12. https://doi.org/10.1038/srep21170. (Nature Publishing Group).

Xia, C., et al. (2019). Evolution and synthesis of carbon dots: From carbon dots to carbonized polymer dots. *Advanced Science*. https://doi.org/10.1002/advs.201901316. (John Wiley and Sons Inc).

Yan, Y., et al. (2019). Recent advances on graphene quantum dots: From chemistry and physics to applications. *Advanced Materials, 31*(21), 1808283. https://doi.org/10.1002/adma.201808283. (John Wiley & Sons Ltd).